Sex in the ~~fu~~

Dr Robin Baker was Reader in Zoology in the School of Biological Sciences at the University of Manchester from 1980–1996. Since leaving academic life in 1996, he has concentrated on his career in writing, lecturing and broadcasting. He has published over one hundred scientific papers and many books. These include the international bestseller *Sperm Wars*, which was based on his own original research on human sexuality, and *Baby Wars*, which exposed the conflicts at the heart of even the happiest families. His work and ideas on the evolution of human behaviour have been featured in many television and radio programmes around the world. He has five children and has lived in Manchester since 1974.

Also by Robin Baker in Pan Books

Sperm Wars

ROBIN BAKER

Sex in the Future

Ancient Urges Meet Future Technology

PAN BOOKS

First published 1999 by Macmillan

This edition published 2000 by Pan Books
an imprint of Macmillan Publishers Ltd
25 Eccleston Place, London SW1W 9NF
Basingstoke and Oxford
Associated companies throughout the world
www.macmillan.co.uk

ISBN 0 330 37034 0

1 3 5 7 9 8 6 4 2

A CIP catalogue record for this book is available from
the British Library.

Typeset by Intype London Ltd
Printed and bound in Great Britain by
Mackays of Chatham plc, Chatham, Kent

To Liz – with thanks for the warning

Contents

Part Four: Relationships to Fear – or Not to Fear?

9. Commissioning a Family, Infidelity and the World's Population 263

Scene 9: A Couple of Lone Parents / Pedigrees and Families – Forty Years On / Living Together / The Commissioning Parent; Head of the Family / World Population in the Balance / Infidelity – Redefinition or Oblivion?

10. Incest, Relationships and the Law 279

Scene 10: Like Mother, Like Daughter / An Ancient Aversion / The Incest Taboo / Cloning and Relationships / Legal Quagmires

11. Homosexuality, Nucleus Transfer and the Future 301

Scene 11: Round Robin / A Rich and Varied Gene Pool / Bisexuality and Homosexuality / Gamete Manufacture / Homosexuality in the Future

12. Extended Families in the Future 317

Scene 12: Keeping in Touch / Killing the Extended Family, Swamping the Gene Pool – an End to Evolution? / Extended Families: Past, Present and Future / Swamping the Gene Pool / Evolution: Alive, Well and Shaping the Twenty-second Century / And the Winners Are . . .

13. Future of the Oldest Profession 331

Scene 13: Making a Living / Pimps and Prostitutes . . . Come in from the Cold

Part Five: Time-warps

14. Reproduction Past Menopause – and Beyond 341

Scene 14: The Three Musketeers / Errors in Ancient Programmes / Menopause / Reproduction after Death / Reproduction Restaurant – Babies to Go

Acknowledgements

As usual since I left the sanctity of academia to enter the maelstrom of popular science writing, my partner has made an invaluable contribution to the book's contents. In conversation, she saw problems with much that I told her I was about to write. On some occasions, I eventually came to agree with her. On others, I didn't. But even then, the deep discussions we had considerably sharpened my appreciation of what I was trying to say. I have no hesitation in dedicating this book to her.

Sex in the Future has neither footnotes nor references. Yet the book could not have been written without the basic research carried out by huge numbers of people. A sincere acknowledgement, therefore, must go to all those unnamed scientists whose research, insight and technological wizardry have contributed in any way, large or small, to the ideas, stories and procedures presented in these pages. For the sake of any readers who might wish to pursue the science and technology further, I have added a small bibliography.

I thank Kim Cotton of COTS (Childlessness Overcome Through Surrogacy) and Dr Pease of St Mary's Hospital, Manchester, without whose advice I might have slipped up in one of my scenes. My final acknowledgement, though, is more in the form of an apology – to the School of Biological Sciences at the University of Manchester in general and to my two former research students, Chris Bainbridge and Charlie Nicholls, in particular. In August 1997 I decided that the only way I could write this book would be to give up virtually all my lingering university commitments. I thank everybody who helped me do so. I left Chris and Charlie in capable hands and I trust that by the time this book is published, they will both have been awarded their PhDs.

Introduction

Ancient Urges Meet Future Technology

'Ancient urges' – a powerful phrase conjuring up thoughts of wanton promiscuity and public sex. Images form of lustful cavemen dragging subservient females by their hair, or of hedonistic women bestowing liberal sexual favours on ever-potent men. This is not the way most of us run our reproductive affairs – but the beast is in us, none the less.

Who can deny that the human sexual repertoire, from its most basic to its most subtle, is anything other than ancient – etched into our bodies by millennia of natural selection? Each person's anatomical gender is determined the moment their father's sperm enters their mother's egg. Thereafter, their sexual destiny is a roller-coaster of genetically orchestrated development. Men develop a penis and testes, manufacture sperm, have erections, and enthusiastically seek women to inseminate. Women develop a vagina, womb and ovaries, and manufacture eggs. They organize their body chemistry and behaviour around a menstrual cycle, and – albeit with perspicacity – seek men to inseminate them.

We are forced to accept the antiquity of such programming as a matter of both logic and observation. Logic tells us that if humans weren't programmed to reproduce sexually – and to have the body, body chemistry and behaviour with which to do so – then there would be no humans. Observation tells us that our reproduction processes are no different from our nearest relatives, the primates.

Our sexual legacies have served us well for over a million years. But as humankind strides into the twenty-first century, the old animal within us is under threat. How *will* the beast cope with those developments that almost without warning have become part of our reproductive destiny – IVF, cloning, surrogate mothers, surrogate testes, nucleus transfer, gamete banks and frozen embryos – or even with simple paternity testing and child support enforcement? Many people are worried – even terrified – by the potential consequences. So rapid are the scientific achievements that governments, legal systems, theologians, ethicists – and public comprehension – are all being left far behind and breathless. Fears abound that the new technology will create monsters, overturn traditional values, and generate new and undesirable relationships.

'Unnatural', say many people. 'Demeaning', 'a threat to human dignity' or 'contrary to the will of God', say others. Opposition to change is immediate, almost instinctive, and more often than not is fuelled by ignorance of human reproductive biology. One of the themes of this book is that gut reactions to what is 'unnatural' are often both illogical and non-biological. So, too, are claims that the modern technology is an assault on human dignity and individuality.

I shall argue that modern technology has the potential to do more than put an end to infertility. In the process, it can also tip balances and correct ancient inequalities. The result will be the creation of a more even-handed and certainly much more varied sexual arena than has ever existed in the past. At the same time, for good or for bad, it will free much more time for men – and particularly women – to concentrate on things other than reproduction.

The portent of the book is optimistic. The human species will both continue and be enriched – rather than demeaned – by the modern technology. Dignity and individuality will flourish undiminished. When ancient urges meet future technology, there will be no winner or loser. Instead the two will join forces and, with a little help from the immortal power of natural selection, will propel human reproduction through the next 100 years of its evolution with profit not cataclysm. Such is the prediction.

There is another prediction. During the first half of the twenty-first century, the divorce of sex from reproduction – which is currently nisi – is expected to become absolute. The result will be greater sexual freedom and an ever-burgeoning range of *reproductive* choices. The latter will be as varied as the menu of a good restaurant – a Reproduction Restaurant.

At first, only the infertile will benefit from the full menu, but there will be no stemming the demand for equality from the fertile. They, too, will want the same range of privileges. For the fertile, conception via intercourse will be on the menu, of course, but it will be one of the cheaper options. For wealthy gourmets, whether fertile or not, there will be a multitude of other possibilities, needing careful thought before reaching a decision. Should a couple use sperm and eggs to reproduce? If so, should they be freshly collected or banked? Or should they use manufactured gametes, derived from cell lines they banked at puberty? Should they reproduce with somebody they know – a joint venture? Or should they go it alone and purchase the gametes of somebody famous – or even dead? And if they decide to go it alone, maybe they should use one of their own diploid cells and so parent a clone? Should they reproduce with somebody of the same sex as themselves or of the opposite sex? If they are female, should they gestate the child themselves or should they commission a surrogate? And when should they have their first child – in their teens, or in their twenties, thirties, forties, fifties – or even sixties? The range of choices will be almost endless, the Restaurant menu complex, but time, advice and counselling will abound.

There is more to this than simply range of choice, however. In the United States donor sperm and eggs can already be obtained with relative ease and couples living elsewhere are taking advantage of the fact. Some services even combine express-mail delivery of eggs and sperm with World Wide Web sites that allow prospective parents to screen for donor characteristics such as height, weight and eye colour. Assisted reproduction tech-nology is already a growth industry in the United States thanks to the market forces of supply and demand. In the future, most such gamete selection will take place over the Internet. Just as the twentieth century saw the formation of Internet Cafés, so too could the twenty-first century literally see the establishment of Reproduction Restaurants, places where people could go to eat, drink and browse their reproductive possibilities – and maybe even commission a child over a gourmet meal and a bottle of wine.

Sex in the Future, just like my two previous forays into popular science, is a collection of short stories interspersed with scientific discussion. Here, the aim of the scenes is to give a flavour of reproductive life in the not-too-distant future and to show that when seen through the eyes of characters rather than through the stark imagery of scientific jargon, nothing seems quite as scary.

While this book was being written, the first trickle of publi-cations on cloning and other assisted reproduction technologies was beginning and I daresay that by the time it is published, that trickle will have become a flood. The focus of *Sex in the Future*, however, is different from the other literature I have seen so far. My interest is in the *behaviour* of the people who will live through the twenty-first century. I am less concerned with the details of technology, the fears of the ethicists, tribulations of the theologians and principles of the politicians. These are all discussed, of course, but it is people's behaviour that takes centre-stage. That is why the book begins, not with the technology of assisted reproduction, but with paternity testing and child support legislation. These will have just as important an influence on babymaking in the twenty-first century as will IVF, surrogacy and cloning.

The book is organized in roughly – but by no means precisely – chronological order, and after each scene I have tentatively suggested a year in which the action might be set. These dates, like the settings, are focused primarily on the major industrial nations. Although I have little doubt that eventually the people of the Third World will tread the same path – as now with contraception – their road to the Reproduction Restaurant will inevitably be longer, slower and more tortuous. But arrival is only a matter of time.

Towards the end of the book, several scenes show the Reproduction Restaurant in action. After this Introduction, though, the first scene begins more or less now – at the dawn of the twenty-first century.

Part One

Decline of the Nuclear Family

1

Paternity Testing and Child Support

SCENE 1 (VERSION 1)

My Son – Perhaps?

The phone rang.

Drunk as usual at this time of the evening, Jim tried to ignore the persuasive sound. With only minutes to go to the end of the match he was watching on television, he was reluctant to miss the dénouement. The result was only of academic interest to him, but he knew that whoever was on the phone would be of even less interest.

The ringing was both persistent and insistent. Eventually, Jim pulled himself to his feet and with a slight stagger walked into the hall, clutching his nearly empty can of beer. Picking up the once-white phone he tried to say hello but the phlegm in his throat muffled his words. It was hours since he had spoken. He cleared his throat but as he inhaled, tiny flecks of cigarette ash from the receiver were sucked into his throat, making him cough.

The operator asked Jim to pay for the call. Taken aback, he reluctantly agreed.

'Hi, Dad,' came the unexpected voice. 'You sound rough.'

Jim was genuinely pleased to hear his son's voice. It was at least two months since they had talked – and they'd ended on bad terms. They always ended on bad terms. Between times, Jim would fret – and vow to be more tolerant and amenable next time. But it wasn't easy, especially as the gaps between conversations were so long. The pair lived 200 miles apart and although Jim *would* make contact more often, there was no phone in the condemned house his son called home – so all Jim could do was wait.

The caller was Rob, the youngest of Jim's three children. Dan, his eldest, was nearly thirty, ran a successful business and now had two children of his own. Sarah was in her mid-twenties, had just finished a PhD and had one child. Then there was Rob – just twenty and sending Jim to an early grave with worry and penury.

There was no doubt of Rob's innate talent. An aspiring poet, he had shone at school and easily earned a place at university – but his student career had lasted just two months. Too lazy and hung-over each morning to attend his classes and too misguided to heed his tutors' warnings, he had assumed he was an asset that couldn't be thrown off his course. He was wrong but didn't admit his dismissal to Jim until the end of the year when he could hide it no longer. In the meantime he had continued to spend Jim's money – as well as building up overdrafts and acquiring loans. He claimed he had been waiting for inspiration.

It was nearly a year now since the big argument when Jim had told Rob he was on his own – that if he couldn't take advantage of the opportunities that were handed on a plate then he would have to make his own way. Rob had carried on living in his univer-sity town, spending his nights – and much of his days – sleeping on floors offered by sympathetic friends who were still students. Mainly he tried to live off welfare, supplemented by a succession of menial cash-in-hand jobs. These ranged from packing boxes of chocolates to working in bars. His latest job was stuffing chickens. He had written a few poems, but unimpressed editors had returned them all.

The phone call began reasonably well. Rob said he was happy – more or less. And he was healthy, apart from an outbreak of zits. When Jim asked, Rob eventually admitted he'd lost his job stuffing chickens. He claimed he'd had a couple of girlfriends since they last spoke, but neither had lasted long. After a while, Rob broached the reason for his call – and to Jim, it had an all-too-familiar ring to it.

'Look, Dad. The reason I'm phoning . . . Well, I've had this letter from the bank. Well, not really from the bank – from debt collectors. They say I don't have a bank account any more. It's been handed over to these debt guys or something. Anyway, they say if I don't pay off my overdraft, I'm going to be blacklisted – or something – *and* they're going to take legal proceedings. They wouldn't do that, would they?'

Jim's heart sank. 'Rob, I told you never to ask me for money again. I told you to sort yourself out. I haven't got any money myself, thanks to you lot. Ask your mother for some money – *ask your brother* – but don't ask me. I've got enough problems making ends meet myself, thank you.'

That didn't end the conversation, but it heralded the end. Rob persisted with his request, eventually pleading tearfully for Jim to pay off his debts. He wouldn't let it happen again, he said. He promised. He really meant it this time, he said. In the end, Jim angrily agreed at least to think about it. Then he slammed down the phone, all vows of tolerance and amenability shattered again. Part of the problem was that Rob could only remember the old days, when Jim had part-owned his own business, was well off and generous to a fault.

Exasperated, Jim downed the remainder of his beer in one gulp and returned to his chair in front of the television. He was disappointed that Rob had not simply wanted to talk but he was also angry at the clumsy attempt at emotional blackmail and the false promises. Then he felt panic, first at the thought of what *not* helping might do to Rob, then at the thought of what helping might do to himself – even if he could somehow scratch the money together. More than anything, though, he felt guilty – because the conversation had raised in him the old, old suspicion. Could Rob really be his son?

The doubts had been there from the very beginning. Two children were enough, Thelma had said. We've got our family – a boy and a girl – let's concentrate on them. Let's do everything we can for *them*. But then, out of the blue one evening, when their marriage was becoming increasingly shaky, Thelma appeared before him, naked from her bath, proclaiming she wanted another baby. A month later, she was in the throes of pregnancy sickness. For Jim, the whole episode had the aura of deception. And although Rob's birth and infancy kept them together for a further seven years, he could never convince himself that he was the boy's father. Eventually, he and Thelma separated and divorced.

Jim glanced at the clock – 21:50. Time to sober up. Rising from his chair, he stared out of the window for a few moments at the dark, windswept streets. The last thing he felt like doing was to go out – but he had no choice. How on earth had he ended up in this state, he asked himself. It was a rhetorical question, though – he knew the answer.

On their divorce, Thelma kept the marital home and Jim, beginning what should have been a brief climb back to wealth and home-ownership, rented a sizeable, almost flamboyant, flat. But from then on, he misjudged everything.

Thelma never did cohabit with anyone; nor did she get a job, preferring to live off Jim's maintenance. Dan went to university, spent Jim's money as if it was infinite, and a year after graduating begged for capital to buy into a friend's family business. Sarah followed Dan to university, succeeded in being an even greater financial drain, and partied herself into pregnancy. None of the three potential fathers would accept responsibility, so Jim ended up paying for good accommodation and child-care while Sarah finished her degree.

Jim raised the money Dan and Sarah needed by selling to his oldest friend, Dave, his half of their joint business. In the years following their graduation, Jim and Dave had built up a successful advertising company. Reluctantly, Jim sold his business partner everything on the understanding that his job and privileges would stay secure. He then used the capital he had raised in two ways. He helped his two eldest children – and he bought a house to placate his new girlfriend.

Susan had been more than happy to visit Jim in his flat in the early days. But as soon as her place in his life seemed secure she pressurized him into buying a property that reflected the status *she* wanted from the relationship rather than the status Jim could afford. Twenty years younger than Jim, Susan was everything he had been looking for in a second wife, even though deep down he knew that she was mainly after his money. They married and bought the house in joint names, even though all the capital was his.

It soon became obvious that Jim's income could not support their lifestyle and before long there were bitter arguments between them. His financial demise was the image in reverse of his son Dan's financial ascent, as Susan never hesitated to point out. Almost a year to the day after Jim and Susan had moved into their new house, Jim came home unexpectedly from what should have been an overnight business trip to find Susan in bed – with Dan.

The bitter and abusive scene that had followed his interrupting Susan and Dan's lovemaking was the end of his relationship with both of them. Susan moved out to live with Dan, and as part of the divorce settlement took half of the equity from their house.

Soon after, to Jim's amazement and despite his opposition, Dave and Dan merged their businesses and became partners. Dave had always taken an interest in Dan and had helped him out in small ways. 'Uncle Dave', Dan had called him when he was young. But for Jim, this was the end. Feeling unable to work for the team of his ex-business partner and his son, Jim left the company – only to find that men over fifty, no matter how successful their past life had been, could not find employment that easily. In the end, he had been forced to move into the tiny flat that was now his home. He could just about make ends meet but had nothing to spare to pay off the debts of idle young poets.

Leaning over the bathroom sink, Jim splashed cold water over his face. He didn't need to be totally sober for what he had to do – but he needed to be more sober than he was. After studying his face in the mirror, he decided he didn't need to shave. Who was going to see him, anyway? Who ever saw him?

As he wandered around his tiny kitchen, making sandwiches and flasks of coffee – and around his bedroom, donning his night security guard uniform – he pondered his younger son's paternity for the thousandth time. None of his children looked particularly like him, resembling their mother much more closely. But at least the two eldest had his drive and sense of purpose. He couldn't see anything in Rob's appearance, personality or demeanour that helped to convince Jim that he was his son. And his conception – so unexpected and so deliberate – seemed just one more act of deceit. Had Thelma really wanted them to have a third child – or had she been unfaithful to him and simply been trying to cover her tracks? Yet here Jim was, twenty years on, being asked for what little money he had – and for what? To save a lazy, useless adolescent – who was probably somebody else's child – from ending up on the streets or in jail.

What irritated him more than anything was that both Rob's mother and his elder brother were better placed to help Rob than he was – but both were under the thumb of hawkish partners. As Jim closed his front door and stepped out into the dark street, he made his decision. This time he *would not* pay off his 'son's' over-draft – and as if to emphasize his decision, a train thundered over the nearby railway bridge.

*

On the other side of the railway track, both geographically and socially, Jim's two ex-wives were finishing an expensive meal.

Thelma was very drunk and feeling very affectionate towards her daughter-in-law.

Theirs had been a relationship of two halves. At first Thelma had resented her place as Jim's wife being taken by a girl young enough to be her daughter. But when Susan metamorphosed from second wife to daughter-in-law, then gave Thelma two beautiful grandchildren, their relationship had blossomed. Even the last tinge of jealousy – because Susan was actually the wealthier of the two – had disappeared when Thelma had finally found somebody suitably wealthy and had married again.

'There's something I've always wanted to ask you,' said Susan as they both sipped liqueurs. She knew her mother-in-law was very drunk and sensed that tonight might just be the night to ask. 'Do you know the one thing that really screws him up?'

They always referred to their mutual ex-husband as 'him'.

Thelma laughed, swaying slightly in her seat. Her cheeks were flushed and her eyes bloodshot. 'I know – you don't need to tell me – I know what it is. He'd like to think that that lazy good-for-nothing son of ours isn't his. Wouldn't he? That's what he'd like – so he doesn't have to give him any more money. I know just what he's thinking.'

Susan nodded, mildly surprised. 'Except, he reckons he's always had his doubts. Even when you were pregnant.'

'I know. He used to keep asking me. About once a month he'd ask me. In the end, it was one of the reasons I left him. I kept telling him Rob was his, but he never really believed me. He got totally screwed up about it.'

Susan paused a second, but couldn't leave it there. 'So go on then . . . is he? Really, come on, tell me . . . is Rob really his son?'

Thelma smiled, took another sip of her liqueur, then leaned across to whisper her secret. She blinked like an owl, as if she was going to fall asleep on the table. Susan tilted forward until their faces were so close she could smell her mother-in-law's breath.

'Yes, he is,' Thelma said, with emphasis. 'He really is his son . . .

but I *will* tell you something – as long as you promise to keep it to yourself.' She paused for agreement and effect. Gaining both, she carried on. 'They might not *all* be his.'

This was the sort of divulgence Susan wanted. 'You mean . . . Sarah's not his daughter?' she said in amazement.

The elder woman shook her head. 'No,' she said. 'It's not Sarah.'

'You mean . . .'

Thelma laughed, enjoying the impact of her revelation. 'But for goodness' sake don't tell anyone. I can't be *absolutely* certain. Jim *could* be his father – it's possible – but if any of them isn't his . . . it's Dan!'

Taming the Forces

A divorced ex-businessman is down on his luck. Money-grabbing ex-wives share the secrets of their infidelities – one of which was with his eldest son. A promiscuous daughter falls pregnant. An idle son bleeds his father for every penny he can. Such a depressing scene hardly mirrors the fate of every man, but it is common enough to strike a chord with many an observer at the end of the twentieth century. Why was the man in this position? And would he be better or worse off in the future?

Most of the chapters in this book will be concerned with the impact of modern technology on the nuts and bolts of human reproduction – the micro-manipulation of testes and ovaries, sperm and eggs, cell nuclei and the like. The two chapters in Part 1, though, are different. The legacies under review here are subconscious urges that control what is, in effect, the *strategic behaviour* of men and women. Equally as innate as their more anatomical counterparts, these legacies surface as each sex tries to do the best, reproductively, for itself and its children. The impacts of modern developments are much less tangible than the more obvious impacts of technology on conception and fertility – but they are no less potent, as we shall see.

In the scene, the main characters were struggling against – or taking advantage of – behaviour patterns that have been handed down from distant ancestors. The main forces at work, sculpting the scene's events, were the influence of paternal certainty on men's behaviour and the problems and opportunities that this generates for women. Soon, though, these forces will be tamed – and the first faltering steps in two different directions have already been taken.

At inception, these two moves – paternity testing and child support enforcement – were expected to bolster the institution of the nuclear family, which by the end of the twentieth century was in free fall. Ironically, as we shall discuss in Chapter 2, they are likely to have exactly the opposite effect. This is because, socially desirable though both are, they actually erode the cement that binds the nuclear family together – the threat of infidelity.

First, though, this chapter considers the biological background, social birth and initial influence of both developments.

Paternal Uncertainty and Paternity Testing

Paternal Uncertainty

There is a much underestimated psychological difference between men and women. A woman can have total confidence that she is the genetic parent of any child to whom she gives birth. A man can *never* have total confidence in his genetic parenthood. Sex takes only a few minutes, so no matter how much of his time a man spends with a woman – unless it is absolutely every minute – a nagging doubt over paternity can always remain. Trust helps, but can never be total. And the less a man trusts a woman, and the less time he spends with her, the greater that doubt can be.

A survey in the UK in 1989 showed that if a man spent more than 80 per cent of his time with his partner between sexual acts, she was almost never unfaithful to him. Less time than that, though, and the chances of her infidelity increased significantly, rising to over 10 per cent if he spent less than 10 per cent of his time with her. Other facts also emerged, such as that a woman was most likely to have sex with a lover during the fertile phase of her menstrual cycle and that acts of infidelity are less likely than routine acts of sex with a live-in partner to involve contraception. The result is that from time to time, infidelity makes babies. On average, the supposed father does not sire 'his' child on about 10 per cent of occasions – men's worries over paternity are sometimes justified.

In the scene, Jim spent much of his life agonizing over the paternity of his third child, Rob. And the more he was required to do for Rob, the greater his preoccupation. He fretted that any parental effort on his part was misdirected – if Rob was not his. The feeling that Thelma and some unknown paramour might have tricked him gnawed away at him constantly. And like most men in his position, Jim's uncertainty affected how good a father he was to his child.

In one sense, though, Jim was unusual. Most men spend little time *consciously* pondering their paternity. But this is relatively unimportant. Whether a man's uncertainty is conscious or subconscious, the quality of care and support he shows for a child is higher if that child is his genetic offspring than if it is not – as study after study has shown.

Quality of Care

Most such research has examined the behaviour of men in blended families – households in which some of the children are the man's genetic offspring and some are his stepchildren. Studies of families in Canada and Britain show that a man in a blended family is seven times more likely to abuse his stepchildren than his genetic children – and a massive *100 times* more likely to kill them. This increased risk of violence and murder on the part of stepfathers is greatest for young babies up to the age of two, but is not confined to that age group.

There is nothing particularly new and twentieth-century about this behaviour. Studies of the forest-dwelling Ache people in Paraguay, for example, showed that 9 per cent of children raised by a mother and stepfather were killed before their fifteenth birthday compared with less than 1 per cent of those raised by two genetic parents.

Abuse, neglect and murder are the most extreme forms of parental disfavouritism and fortunately are relatively rare. Much more common is simple parental favouritism, in which one child gets more toys, more pocket money and simply more praise and attention than another does. Again, we should not be surprised to find that in blended families men are more likely to favour their genetic offspring than the offspring of another man.

Humans are not the only species to care about genetic parenthood. Male birds spend more time helping to feed the chicks of faithful than unfaithful partners. Lions are much more likely to kill cubs that cannot be theirs. Monkeys and apes also show heightened aggression towards youngsters that cannot be theirs. We can never know, of course, how conscious or unconscious is the behaviour of the males of these other species – and it really doesn't matter. Probability of paternity affects their behaviour.

And the youngsters who masquerade as their offspring benefit or suffer accordingly, as do the youngsters' mothers.

Evidently, the male behaviour associated with paternal uncertainty has a long evolutionary history. Perversely, though, this traditional problem for men is equally a traditional *opportunity* for women. Females can sometimes gain from male confusion over paternity – as Thelma did in the scene.

Confusing Paternity

If a woman can convince her partner that he is the father of her child when he isn't, then she gains and her partner loses. In the scene, Thelma seemed to have engineered just such an advantage for herself in conceiving Dan. Jim poured money into Dan's education and career, little realizing that he might not be his son.

The converse can also happen – a woman failing to convince a man of his paternity even when he really is the father. In the scene, this is what happened at Rob's conception. And when things became tight, Jim discriminated against Rob as a result.

As these two fictional conceptions illustrate, the pressure has always been on the man to judge the situation correctly and on the woman to convince her partner that he is the genetic father, whether he is or not. Sometimes one or other succeeds and sometimes they don't. In the scene, Jim failed miserably, judging the situation correctly with only one – Sarah – of his three children. Thelma succeeded a little better, convincing Jim of his paternity of two of her three children, despite the fact that Dan might not have been his.

Thelma also achieved something else – but to explain it requires a secret to be divulged. At the end of this chapter we revisit Jim, Thelma and their children and consider what a difference it would have made to their lives if there were no such thing as paternal uncertainty. There we discover that Dan *wasn't* Jim's child. His genetic father was actually Jim's erstwhile business partner, Dave.

Thelma's achievement is emulated frequently by female primates, and is also often contrived by female birds. What each of these females manages is to convince one male that he *is* the father of her offspring while at the same time making one or more other males think that they *might be* the father. Then all the males are much more inclined to help than hinder. Some female birds have been observed to recruit three or four males to help feed their fledglings by this ruse. Female lions and monkeys reduce male aggression towards their offspring by giving several males cause to think that they might be its father.

The way that all these females confuse paternity, of course, is by having sex with more than one male. In the twentieth-century industrial society in which Thelma lived such promiscuity was fairly covert and secretive. In some human societies, though, female promiscuity is much more common and overt. In the Ache of Paraguay, for example, women regularly have sex with several men – and paternal confusion is the rule rather than the exception. So much so that a man who has had sex with a child's mother can categorize himself as either the child's primary father – in other words the man *most likely* to be the child's father – or its secondary father, or even its tertiary father. To differing degrees, all the 'fathers' help in the child's upbringing, to the benefit of both the woman and the child.

In Thelma's case, two different men were encouraged to think they were Dan's father. Jim gave Dan his greatest help, but 'Uncle' Dave also helped him on and off during his childhood. And he did so again, later, when they merged businesses. Almost as if they were Ache, Thelma and Dan both did very nicely out of the situation Thelma's promiscuity had created.

As we move into the twenty-first century, though, life is about to change for the Thelmas, Daves and Jims of the world – and maybe eventually for the Ache. Global paternity testing is just around the corner.

Paternity Testing

The men of the future will be liberated from the spectre of paternal uncertainty that haunted their male ancestors. In the past, a man's only window on paternity was a set of vague recollections of sexual activity around the supposed time of a child's conception, perhaps bolstered by an equally vague physical resemblance to the child. On that flimsy basis, men sacrificed their time, effort and resources in the raising of supposed children. Now, every man can confirm his paternity if he so wishes – and in the future it will probably be done for him as a matter of routine.

Men will owe this liberation to paternity testing, a technological offshoot from the process of genetic fingerprinting. This latter technique was developed in Britain in 1984 by a man – Alec Jeffreys – whose gender is relevant later on in this chapter. We shall discuss the nature of genes and chromosomes in more detail in Chapter 8. For the moment, it is enough to know that genes are simply sequences of chemicals. They are arranged along the chromosomes found in the nucleus of nearly every cell in the body. Chromosomes are composed of a chemical known as DNA (deoxyribonucleic acid). DNA itself is made up of different types of amino acids arranged in unique sequences. Short sequences of amino acids are called bases, and a gene consists of a certain number of bases.

Jeffreys found a way of identifying bits of DNA – called mini-satellites – that flanked particular genes. He also found that everyone has repeated mini-satellite regions in their DNA, but that the number of repeats is unique to every individual. Moreover, these repeats are inherited from the parents, half from the mother and half from the father. Only identical twins end up with the same numbers of mini-satellite sequences. This discovery was destined to change family life for ever – because it heralds the end of paternal uncertainty.

A sample of any tissue that contains cells – blood, saliva, hair roots or semen, for example – is all that is needed to make a genetic fingerprint. DNA is extracted from the sample, then cut chemically at specific points using so-called restriction enzymes. The fragments of DNA produced are placed on a gel and separated from each other via electrophoresis – the running of an electric current through the gel. Pieces of DNA have a negative charge. So when a positively charged electrode is placed at the other end of the gel, the charged fragments of DNA are attracted through the gel towards it. Shorter, lighter fragments move faster and hence travel further while the longer, heavier fragments move more slowly and hence travel less far. The result is that the fragments separate along the length of the gel according to their size. At first, the pattern of fragments along the gel cannot be seen but after treatment, a series of bands becomes visible. This is the genetic fingerprint – a mixture of light and dark bands, darker bands containing more DNA fragments.

The genetic fingerprint looks a bit like the bar code found on goods in large shops and supermarkets. And just like bar codes, it can also be described as a series of numbers, each number referring to the length of the DNA fragment. This length can be calculated by measuring how far a fragment has travelled through the gel compared with marker bands of DNA of known length.

In paternity testing, cells from the child, the mother and the putative father are used to produce a genetic fingerprint for each. As a child inherits half its DNA from its mother and half from its father, these fingerprints can be used to check paternity. Any band in a child's DNA fingerprint that fails to match a band in its mother's DNA fingerprint must have been inherited from its father. If the putative father is the real father, these non-maternal bands from the child will all match with bands from this man.

This sounds good – but the measured match between finger-prints from a child and a putative father is statistical rather than certain. Even when a child's fingerprint is compared with that from a man who is *not* its genetic father some bands (25 per cent, on average) will match by chance. This might sound like a fairly basic flaw in the process, but it is not – as long as enough bands are available for comparison. For example, if both the child and the putative father have, say, ten bands for checking in their DNA fingerprints, the probability of all ten matching by simple chance is very low – approximately once in a million tests. Usually many more than ten bands are available for analysis and a chance match is even less likely. In current usage, paternity testing really is a powerful and accurate procedure.

Paternity tests are a lot easier and a lot more reliable if tissue can be obtained from everybody concerned – the mother, the child and the range of putative fathers. There are ways around not having tissue from the mother, especially if tissue is available from a second child, but the technique becomes a little less reliable. This need for cooperation raises a problem because one or other of the parties may consider that it is not in his or her interests to do so. The question then becomes – under what circumstances should participation be coerced?

This question is in fact only part of a much bigger one. How far down the road towards automatic or enforced paternity testing should governments go? Although this may not be a major concern at the moment, it will become a concern for the future. We shall discuss this question shortly, after considering a related phenomenon – child support legislation.

Deadbeat Dads and Child Support Enforcement

In the scene, Thelma gained from having sex with two men around the time of Dan's conception. She confused paternity and Dan reaped the benefits. Her daughter Sarah, though, who had sex with three men around the time of her own child's conception, experienced a different pay-off. She fell victim to one of *man*-kind's least pleasant but most enduring and perfectly natural attributes – a taste for having sex with a woman and then abdicating all further responsibility. It is this unpleasant but immutable trait which child support legislation is intended to counteract.

Urgent and Irresponsible Males, Coy and Caring Females

There is a basic difference between the sexes, a difference that biologists can trace back over 800 million years to the earliest of multi-cell animals and the very origin of males and females. Sexually, males are urgent and relatively indiscriminate whereas females are much more coy and cautious. It is a truly ancient characteristic handed down over aeons of time – to be as much a feature of men and women as it is of male and female butterflies, drakes and ducks, dogs and bitches, and bulls and cows.

Men have a much more cavalier approach to one-night stands and other casual acts of sex than women have – because men have more to gain and less to lose from such encounters. Their gain is the chance of cost-free reproduction. Men's bodies could produce innumerable children, but they are limited by opportunity. So they take advantage of as many opportunities as they can, no matter how fleeting. After all, what have they got to lose? Not for men the drain of pregnancy, birth and lactation that has bedevilled female mammals from the very beginning. For women, it is different. It is their bodies, not their sexual opportunities, which limit their reproduction. And unlike men, they do have much to lose.

We can add to this basic gender difference in the pay-off for casual sex the fact that if a child results, the man cannot be certain it is his. It is not surprising, therefore, that natural selection has shaped men biologically to be predisposed to 'have sex and run'.

Deadbeat Dads

This ancient male predisposition to have sex then flee creates a problem for a woman. For her, in effect, it turns sex into a gamble. She can't recruit a man to help her raise offspring without at some point allowing him to have sex with her. But having had sex with him, she has no guarantee that he won't then simply disappear, leaving her with the baby. However, although the gamble of sex has been an ever-present problem for the individual woman, it has not been a major *social* problem until the last few decades.

Essentially, a woman deserted by a man has three courses of action open to her. She can try to raise the child single-handedly. She can turn to her extended family for help. Or she can abort, kill or abandon the baby – a social problem, of course, but of a different ilk. In the past – and in cultures such as the Ache now – women did everything but raise the child single-handedly. The second half of the twentieth century, though, has seen an increase in pressure on women to try to raise children on their own. Lone parenthood is on the increase. We shall discuss this important trend in more detail in Chapters 2 and 9, but a few details are needed here to set the scene.

A steep rise in the numbers of lone parents began in the 1960s. By the first half of the 1990s, lone mothers and their children comprised about one in five of families with dependent children in both Britain and the United States. This represented a virtually three-fold increase since 1970. The rise has been particularly steep since 1987, increasing in the United States from 14 to 23 per cent in seven years.

Not surprisingly, lone-mother families on average have less income than two-parent families. In all industrialized nations for which 1990s information is available, children in lone-mother families are at greater risk of poverty. In Australia, Canada and the United States, over 50 per cent of children in lone-mother families are living below the poverty line. In Australia, Norway and the United States, such children account for over half the children in poverty.

For many lone mothers, the harsh economics of lone parenthood have serious consequences for their and their children's survival, health and fertility. There are also social consequences. Lone-parent families *on average* produce children with poorer school performance and higher rates of delinquency. They are also associated with a decline in the mental health of both mother and children. The daughters of lone mothers are more likely to conceive in their early-teenage years, often thereby creating a new generation of such families.

In the last quarter of the twentieth century, lone-parent families formed a societal presence which most governments were eventually forced to address – with varying degrees of enthusiasm and success. The greatest of both was perhaps in Scandinavia. Countries such as Denmark, Finland and Sweden have a high percentage of children in lone-mother families, yet fewer than 10 per cent live below the poverty line, thanks to the mitigating effect of government support.

Even with governmental enthusiasm, the inescapable fact was that lone motherhood was becoming a very expensive phenomenon. Throughout the industrialized world, regimes sought ways to reduce the cost of lone mothers to the state – and quite reasonably they focused on ways of unburdening state responsibility on to the children's absent fathers.

In particular, government knives were drawn on the so-called 'deadbeat dads' – fathers who not only removed themselves from any form of child support but who often covered their tracks so successfully that they could not be traced. Suddenly, these men were being blamed for a variety of social ills from poverty and social pathology to the spiralling costs of welfare – and the statistics were impressive. For example, in the United States 62 per cent of custodial mothers received no support from the child's father – which represented over 6 million lone mothers managing on their own.

What a difference it would make to the women, the children, society – and to the Treasury – if somehow men could be forced to support their genetic children whether they lived with them or not.

Child Support Enforcement

At the beginning of the 1990s, the move towards child support enforcement began to roll. Various governments mooted – and their wider societies applauded – vitriolic suggestions such as licence revocations, wage garnishment and even imprisonment for deadbeat dads.

As an example, a system for child maintenance was introduced in Britain in 1993 with the establishment of a government agency called the Child Support Agency (CSA). Under the Child Support Act 1991, child support maintenance was an amount of money that an absent parent – invariably an absent father – was to be forced to pay regularly as a contribution to the financial support of his children. The money was to be paid to the CSA who would then pass it on to the mother. Similar agencies were formed in, for example, Australia, and each state in the United States drew up its own system for handling the situation.

The amount of money a father is required to contribute is calculated by formula. In 1997 the British CSA provided a description of the calculation process as presented to the absent parent. According to this, the CSA allows essential living expenses for the absent parent and any of his own children who live with him, plus rent or mortgage costs for the house he lives in if he or his partner is the householder. In some cases a weekly allowance may also be provided for travel to work and/or for transferred property or capital. The CSA deducts the total of these costs from the absent parent's net income (which means the amount left after paying tax, national insurance contributions and half of any pension contribution). This net income is known as *assessable income*. In broad terms, the CSA assesses maintenance at half of this amount up to a weekly minimum maintenance requirement.

The CSA does not routinely take account of any debts incurred, and cannot adjust the maintenance to help pay other bills first. The agency always double-checks that the amount assessed is not more than 30 per cent of the absent parent's personal net income, and frequently it is less. The CSA takes account of any new family the absent parent may have and *sometimes* reduces the amount of child support payable so that the absent parent is left with a certain level of income.

The financial circumstances of both parents are taken into account. If the parent with care has a steady income, this may reduce the amount of maintenance that the absent parent is required to pay.

Such calculation processes are very similar everywhere, but there are a few variations – such as between states in the United

States. In Maryland, a child support order is based on only the incomes of the two parents. In Delaware, a basic living allowance for the non-custodial parent is permitted before payment of child support is required. This allowance, called a 'self-support reserve', is set at the federal poverty guideline. Similarly in Britain, the CSA's only legal requirement is to leave the non-custodial parent with the welfare minimum on which to live.

The manifesto of the British CSA states: 'The Government believes that children are entitled to the financial support of both parents.' Few people would disagree with such a laudable belief. Nor can there be any doubt that *eventually* governments around the world will act upon the child support principle fairly and efficiently. Sadly, though, the first attempts at enforcement were anything but fair and efficient. Instead they were ill-prepared, pragmatic, draconian and vindictive and did little to win universal sympathy for the underlying principle.

In Britain, the popular concept of the CSA at the time of its creation in the early 1990s was of an organization that would track down deadbeat dads and *make them* support their children. But under pressure from the Treasury for quick results and in the face of the enormous task of actually locating deadbeat dads, the CSA opted to ignore the difficult cases and concentrate on the easy ones. Responsible fathers found their existing court orders for maintenance being torn up by the CSA, to be replaced with much greater assessments. Yet at the same time, deadbeats who had never paid maintenance continued to avoid payment. Men struggling to bring up two families on a low income found that the formula used by the CSA to calculate maintenance left their new families destitute.

No problem, said the CSA – all you have to do is appeal. But so hastily had the agency, and particularly its computer system, been set up that it simply could not handle the volume of appeals it received. While second families starved, distraught men committed suicide and thousands of people demonstrated outside the

Houses of Parliament, the CSA failed to answer the letters sent to it, answered the wrong letters, or sent out contradictory responses. Few attempts at implementing a new bureaucracy can ever have been so badly organized and the whole episode would have been hilarious had the consequences not been so serious.

The so-called appeal process metamorphosed into a safety net for agency inefficiency, not a safety net for hungry families. Months down the road of appeal, many men eventually had their cases rejected because it was deemed that only those relating to misapplication of the CSA's rules were allowed. The fact that a man's new family could not survive on the meagre amount left by the CSA was not considered to be grounds for appeal.

Inefficiency is a relatively easy problem to solve. First, there needs to be a desire to improve. Second, computer systems – and staff – need to be adequate for the task. There are already signs that the initial inefficiencies of child support agencies are beginning to fade; though in Britain at least the process is proving to be painfully slow. Even though the British CSA was set up specifically to reduce the welfare benefits bill, by 1997 – four years after inception – the agency was still managing to pay maintenance to only one in three lone parents on benefits. In addition, a quarter of all its assessments were still thought to be wrong. Assessments were no quicker by 1997 than they were under the pre-CSA court system. The new computer system that should greatly improve efficiency is unlikely to be in place until after the year 2000. Yet by 1997 there was a backlog of 225,000 cases and by the year 2000 the caseload was expected to have more than doubled. Even so, we can hope that by 2010 the system will be working smoothly.

A more difficult but important problem to solve is the widespread feeling that luck and circumstance rather than socially determined policy decide who is targeted for enforcement and who is not. The phenomenon of deadbeat dads may have precipitated child support enforcement, but a few years on it is clear that other factors are at work. Many men feel that the system is unfair and that they are being victimized while others escape. In large part, whether a man is caught or escapes depends on the attitude of his ex-partner.

In the United States, only half of all women who qualify for child support have obtained enforcement orders on the men concerned. The remainder have a variety of reasons for not taking advantage of support programmes. Some (30 per cent) simply do not want child support and have not asked for it. Others (20 per cent) have accepted alternative financial arrangements from their ex-partners rather than get involved with child support enforcement. Yet others (25 per cent) have accepted that the absent father genuinely does not have the money. Of those absent fathers who have been served with an enforcement order, about half comply in full, a quarter comply partially and the remaining quarter fail to comply at all.

Men are much more likely to help support their children if they have access to them. Almost all men (90 per cent) with full joint custody – and most (80 per cent) with at least some visitation privileges – pay their support in full. In contrast, less than half of those with no visitation privileges comply. It seems that if they were so inclined both women and governments could increase compliance rates simply by allowing men more access to their children – and in 1998 in Britain a move to link child support payments to access rights did indeed begin.

Overall, only about 10 per cent of the 6 million American women listed in government records as being lone mothers are the actual victims of deadbeat dads – though at around 600,000 women that's still a big enough social problem. On current success rates, about half of absconded fathers are eventually located. When traced, however, not all the men on record as deadbeat dads are quite what they seem. Government records being government records, some of the men so recorded are found to be still living with the mother. Other dads are found to be not just deadbeat but actually dead.

The only way to free the child support system from its aura of unfairness and victimization is for all parents – married, separated, divorced, cohabiting and never cohabited – to be treated equally. There should be no favouritism in calculating support for children from first and later relationships, or for children that are living with or apart from the parent. Discrimination – generated by the formulae used to calculate support – is the root of the real problem, the one that has caused most of the antagonism and misery.

Part of the objection to the current formulae is the strong suggestion of a punitive element for parents who are not living with their children or who have dared to have a second family. For example, *custodial* parents who are living in poverty are not expected to support their children financially; in fact, they and their children are actually provided with public assistance. In contrast, *non-custodial* parents who are living in poverty *are* expected to support their children financially. Even when awards do not reduce the non-custodial parent to the poverty threshold, the formulae still favour children from early relationships over those from second relationships.

The production of a fair formula will need a great deal of discussion and analysis. One principle, though, should be paramount – each of a person's genetic children should be entitled to an *equal share* of that person's income and resources. A proportion of the person's income should be deducted for child support then divided equally among his or her genetic children. It should not matter how many people the person has children with, or who they are living with. Nor should it matter in what order they were had, except in so far as it relates to the children's ages.

Of necessity, of course, equality should be moderated by children's ages, older children by and large needing more money than younger. Children, therefore, should count not as individuals but as child units. Thus a baby might, say, count as one child unit, a toddler as two child units, a ten-year-old as three, and a teenager as four. Actual weighting should be based on national – or state or county – figures for the relative cost of keeping children of different ages.

Each parent would be registered centrally as having a number of child units. The total would be updated each year as the children age, become too old for support – or each time the parent has a new child. Thus if a person has ten child units, a child worth two units will receive 20 per cent of the proportion of that person's income that is deducted for child support.

The proportion of assessable income to be deducted for child support will need to be decided as it is for income tax. In fact, *child tax* would be an appropriate name. Presumably, one principle should be that the more child units a person has, the greater *the proportion* of his or her income should be deducted in child tax – just as the more money a person earns, the greater the proportion taken by the Inland Revenue.

To be fair, child tax will need to be collected from *both* mother and father if both are earning, irrespective of which parent lives with the child – again in the same way as income tax. This move has already begun, though out of pragmatism not principle. In Britain, for example, the CSA is using DEOs (Deduction from Earning Orders) increasingly often – in 1997 DEOs were used in 60,000 cases, a more than twenty-fold increase over 1994.

On such a scheme, each person will know – as they do with income tax – that they and everybody else will pay a certain amount of child tax throughout their parental life. Whether and how much they pay will not depend on luck and circumstance as now. Instead it will depend solely on assessable income and total number of child units.

Each child's 'income' will derive in part from its genetic father and in part from its genetic mother, depending on the parents' respective incomes. The total 'income' for that child will not, of course, be paid to the child itself but to one of the parents – usually the main custodial parent – irrespective of what proportion of its time the child spends with each parent. And the parent who receives the money will be responsible for *all* the child's expenses, even when staying with the other parent, save for any voluntary excess contributions made by the other parent. Of course, this means that the receiving parent will be getting back money that they once paid in child tax. But, in a sense, this is what already happens with child benefit – or at least it does if the parent pays income tax or national insurance.

Such a child tax system would be even-handed and non-punitive. Everybody would be taxed in the same way, whether they lived with their child's other parent or not and whether they had children with one, two or more people. There would still be contentious issues, of course – such as what debts and expenses to allow in the calculation of assessable income. But such calculations are a matter of routine for the Inland Revenue and in principle need be no more problematic for child support agencies.

The question will also need to be addressed of how to support children – and parents – when the formula leaves one or other with insufficient money. Again, though, this is not a new problem and need not be exacerbated by child support enforcement. The potential exists to create a fairer system than was ever achieved by leaving people to their own devices.

The final, but crucial, question that child support enforcement needs to address is how to make sure that men are never forced to support children that are not their genetic offspring. This does not mean that men – or women – should be barred from voluntarily adding a non-genetic child to their list of child units, as in adoption. It means only that a man should have the right to know a child's parentage before he agrees to pay tax for that child.

Some form of marriage between child support and paternity testing seems essential.

A Marriage: Paternity Testing and Child Support

To a biologist, one of the fascinating and unexpected consequences of the recent campaign against absent fathers was the way that it flushed paternal uncertainty into the open. Suddenly, societies seemed to be full of men expressing doubts about the paternity of children that they were being forced to support. Everywhere, enforcement agencies had to take seriously the question of paternal uncertainty. In Britain, the CSA made the following offer: 'If you believe you are not the parent you should say so and we may take steps to resolve this ... You may be offered a DNA test at a discounted rate. If this establishes that you are not the father, the Agency will refund the cost of the DNA test.'

Of the *women* who qualify for child support but seek no enforcement order, only 2 per cent cite uncertainty over their child's paternity as their reason. Of the *men* who resist supporting a child on the grounds of paternal uncertainty, 15 per cent have their doubts verified by DNA fingerprinting.

At the moment, the link between child support and paternity testing is fairly *ad hoc* and ramshackle. Yet everybody – men, women and children – would benefit from a more formal marriage of the two, especially if the vows included child support transforming into child tax and paternity testing becoming automatic. A child would be assured of financial support from *both* genetic parents. A man would be assured that every child he supported was his. And a woman would be assured that if she conceived to a man, his money would help support her child to independence – no matter where the man might run.

Just how quickly down the road towards *automatic* paternity testing different governments will be prepared to go is difficult to judge. But one thing is likely – home kits for paternity testing will only briefly be popular. At the moment, it is possible to purchase a kit, take a few cells from inside the cheek of mother, child and putative father, and then send the samples off to a laboratory for genetic fingerprinting. The cost is relatively high – about £400 in Britain – though not as expensive as a lifetime of supporting some other man's child. But important though paternity is to the individual man, it is even more important to governments. The monitoring of paternity must sooner or later become state-controlled – and home kits will have had their day.

Paternity testing is inconvenient – and expensive – once families disperse. So, too, is the process of handling complaints and appeals. Testing would be done most conveniently, cheaply and efficiently soon after the child was born, along with those routine tests of blood group and genetic diseases that already occur. The mother would recently have given a number of blood and other samples. Cells from the likely father – or fathers – could confirm or decide paternity at the same time. Putative fathers are much more likely to be around at the time of birth than years later. Men and women's child tax records could be updated at the very beginning of their child's life, cutting out the cost of hearing and settling appeals of non-paternity later.

In most cases, a woman should be able to name the man – or men – likely to be the child's father positively enough for paternity testing to be straightforward. Only occasionally – one-night stands with an unknown man or following predatory rape – might a woman not be much help in tracking down the father. Most men named as a father for child tax purposes would volunteer to cooperate – just in case. In fact, many might demand a right to such a test. For everybody's benefit – mother, father, child and Treasury – routine and probably compulsory paternity testing at birth seems a future inevitability.

There is, however, an alternative system to mandatory DNA fingerprinting of named men at a baby's birth. This system would cover all cases but raises questions of civil liberty. Along with social security or national insurance numbers, passports and identity cards, every person could have their genetic fingerprint registered. The test could be done at birth and recorded centrally on a global database. Then, given mother and child's genetic fingerprints, a computer search could easily check the man – or men – named by the mother, even if they were no longer around. In most cases paternity would quickly be assigned. Only if there was no match or the mother could give no information would it be necessary to carry out a wider search to identify possible fathers.

At the moment, DNA databases exist only for the fight against crime. Since about 1990, law enforcement agencies in the United States have been collecting DNA samples from convicted criminals. Their aim is to keep such 'genetic signatures' on file, just as they keep fingerprints. By early 1998, the US national database contained DNA samples from 260,000 people, all with criminal records, with 1000 or more being added each month. Police try to match DNA patterns of blood, hair or semen stains found at the scene of a crime to past offenders. The database achieves between 300 and 500 matches a week, and 80 per cent of these matches result in guilty pleas.

Extending the database scheme to cover the whole population would inevitably meet with opposition, but perhaps not as much as might be imagined – because most people would see there were potential benefits. A possible reduction in serious crime is one. Predatory rapists and murderers might often be deterred by the prospect of almost certain identification. A saving in police

– and innocent suspects' – time is another. People could be eliminated from police inquiries without even knowing their involvement had been mooted.

Some people, albeit reformed ex-criminals, have already realized that there are benefits to being part of such a scheme. For example, the US Forensic Science Service often receives letters from previous offenders who would actually like to go on the database so that they can be quickly cleared when new offences occur. At present, though, there is no provision to include volunteers on DNA databases, or to extend the scheme beyond the criminal population.

Although the advantages of a global DNA database are obvious to the majority of people, the price would be a level of infringement of civil liberty that many people would argue was unacceptable. But can societies of the future really oppose such a scheme when the only practical cost would be an infringement of each person's right to get away with felony? In any case, whatever the implications for crime detection and prevention, if child tax gains wide public support it will almost inexorably push governments down the road to routine genetic fingerprinting and paternity testing at birth.

Paternity testing and child support were first linked in the 1990s in the interests of government finance rather than social reform. But their future together lies in creating a much fairer environment within which people of the future can reproduce. Even so, not everybody will benefit from the future's even-handedness, as we are about to see. Some people – women who can skilfully confuse paternity, men who can successfully cuckold others, and deadbeat dads – will be worse off. All would have found opportunities better suited to their talents in past environments in which their vulnerable prey would have been armed only with their ancient guile. Such exploiters will be curbed, not helped, by future developments – but even they would be hard pushed to argue that the future system is less fair than the past.

Imagine how different things might have been for Jim, Thelma and the other main characters in our opening scene if they had been living not in 1990 but in 2035. This will be a world in which paternal uncertainty no longer exists, and fathers have no choice but to help support their genetic offspring . . .

SCENE 1 (VERSION 2)

Winners and Losers

The phone rang in the distant kitchen.

Drunk as usual at this time of the evening, Jim tried to ignore the persuasive sound. With only minutes to go to the end of the match he was watching on television, he was reluctant to miss the dénouement – his answer-phone could deal with the call.

For a few more seconds, the ringing was both persistent and insistent, then it stopped. He thought the answering machine had cut in but a second or so later he heard his young partner's voice. She appeared in the doorway, clutching the cordless phone. Scarcely twenty, Julie was naked save for a towel wrapped around her head.

'It's your son,' she said, offering him the phone.

'Rob? Tell him I'm watching the match. I'll call him back.'

Julie spoke into the phone then offered it to Jim again. 'He says his news is more important than your match. He's going out to celebrate. Speak to him – he's excited.'

Reluctantly, Jim put down his glass of wine and took the receiver. Julie sat on his knee so that she could hear the conversation, muting the TV as she did so. Jim knew what his son's news would be – he'd arranged it, after all. As he prepared to sound surprised and excited, Julie commandeered his glass and began drinking from it.

'Hi, Dad – guess what?'

'You're drunk,' said his father.

'Too right . . . And I'm going out to get even more drunk. I've got a publisher! I can't believe it. Somebody actually wants to publish my work. I haven't even graduated yet, and I'm having my first book of poems published. Isn't it fantastic?'

'Brilliant,' responded his father, just about managing to sound surprised. 'Well done, I'm really proud of you.'

'I can't believe it. It's like a dream. It's too good to be true.'

'I knew you could do it. Next time you're home, we'll celebrate too. Where are you going tonight? Who's going with you? Are you taking your girlfriend?'

'Both of them!' came back the cocky reply.

While father and son exchanged their parting pleasantries, Julie shouted her congratulations down the mouthpiece. Then, as Jim switched off the phone, she drained his glass.

'Get us some more,' he said. 'We should celebrate as well.'

As he watched her walk towards the kitchen, towel beginning to unwrap from her hair with a loose end dangling over her neck, his pleasure at his son's news was tinged with guilt. As a matter of principle, he would have preferred Rob to make his own way in the world, but had been unable to resist giving him a helping hand. It was he who had suggested Rob should send his manuscript to publishers who just happened to employ his company's services. And it was he who had taken the editor to lunch and slipped into the conversation the possibility of what they could do for each other.

'I need to open a new bottle,' Julie shouted from the kitchen. 'Where's the opener?'

'It's there somewhere,' he replied, sinking back into his chair. The match had just finished and the result pleased him. The evening was shaping up nicely.

At the end of their conversation, Jim had promised Rob that he would phone Thelma, Rob's mother, and tell her the good news. Idly, he considered making the effort now. It was thirteen years since his and Thelma's divorce but they were still good friends and kept in frequent touch, sharing the lives and experiences of their two children. Seventeen years they had lived together – surprising, really, considering the bad start.

He still remembered the shock when the paternity test of Thelma's first child, Dan, had shown that he was not the father. He'd had no idea she'd been unfaithful to him. Evidently it had only happened once – with Dave, his then business partner. Just one moment of passion, she claimed. She said that she couldn't believe Dan was conceived that night. But the test proved otherwise. And when at last the child support machinery had run its course, Dave was registered as Dan's father and supporter.

Jim and Thelma had nearly separated – but didn't. And within the year Thelma was pregnant with her second child, Sarah. And this time, Jim *was* the father, or so she said. But he still didn't fully believe her until the paternity test gave him the proof. He was even more suspicious about Rob's conception. Thelma's sudden desire for another child had seemed so incongruous that Jim was convinced she was covering the tracks of another infidelity. Whether she had been or not, the paternity test showed that their single night of sex that month had indeed hit the mark. Jim was in no doubt that the aspiring poet he had just spoken to on the phone was his son.

'I can't find it. Where did you put the bloody thing?' came an irate voice from the kitchen.

Jim was just about to tell Julie to look harder, when he saw the corkscrew on the table next to him. He must have carried it in absentmindedly after opening the first bottle. Picking it up, he staggered drunkenly into the kitchen and handed it over, apologizing sheepishly.

'Are you cold?' he asked, smiling lecherously and nodding towards her nipples.

She snatched the corkscrew from his hand. 'Go and sit down before you fall down. I'll bring your drink in to you – though you're going to be little use to anybody tonight if you drink much more.'

Returning to his chair, Jim smiled to himself. He liked young women. All three of his partners had been twenty when he first began living with them. He'd only been three years older than Thelma but he had been twenty years older than his second partner, Susan – and he was thirty-three years older than Julie.

He and Thelma had separated amicably when Rob – a young poet in the making even then – was just seven years old. By that time, the success of Jim's business combined with the child support that came into their household from Dave had placed them in a comfortable position. They hardly noticed their financial separation.

A year later, Jim had set up house with Susan. She had been an employee of his and as soon as he came 'on the market' had used both her good looks and quick wits to land the comfortable position of sharing his house – and bed. A year after she moved in they had their first child, a son.

It was when Susan was pregnant with her second child that Jim's daughter, Sarah, announced that she, too, was pregnant. Any one of three men at her university could be the father, Sarah had said. She said no to abortion and thanks to lucky timing had no need to interrupt her studies. Perhaps fortunately for everybody, the paternity test proved the father to be neither of the two penurious students she had bedded, but her course tutor. His child support easily allowed Sarah to afford the child-care she needed to continue her studies. She never lived with the child's father, who was already in a relationship, but did live for several years with one of the young students. With support of different kinds from two men, she rarely called on Jim for financial help.

'Here – and don't spill it.'

Jim took the glass of red wine from Julie, laughing at the way she spoke to him as if to a child. Her authoritative air seemed incongruous with her nakedness – and with the difference in their ages. She sat down on a long pile rug in front of the fire, placed her own glass on the hearth and removed the towel from her head. He watched her closely as she pressed strand after strand of her long hair in the towel, drying herself gently.

It was a year now since he and Julie had first begun having sex – acts of infidelity on his part that racked him with guilt until he discovered that Susan was also being unfaithful. He and Susan had both realized they were drifting apart, she wishing that he was thirty-five again, not fifty-three and he wishing that she was twenty again, not thirty-one. Their final separation had been more acrimonious than his separation from Thelma, largely because he discovered that the target of Susan's infidelity was Dan, Thelma's eldest son, whom he had helped raise for sixteen years.

For a while, after Susan had moved in with Dan, taking her two children with her, Jim really began to feel his age. A month ago, though, he had invited Julie to share his home. She needed little encouragement, and they were still very much in sexual mode. He felt young again.

'Come and comb my hair,' she said, 'if you're not too drunk.'

He hauled himself to his feet and went over to her.

'And for goodness' sake, take off those clothes.'

He obeyed, then took the comb from her hand and moved to kneel behind her.

'Don't be silly,' she said, lying back on the rug. 'I didn't mean *that* hair.'

*

A few miles away, Jim's two ex-wives were finishing an expensive meal. Thelma was very drunk and feeling very affectionate towards her son's new partner.

The warmth of their relationship was very recent – less than a year. Thelma had resented her place being taken by a girl young enough to be her daughter and for a while there had been animosity between them. Even when Dan had confided to her that he was sleeping with Susan, she had felt ambivalent. Two recent events, though, had united them. Susan was now pregnant – carrying Thelma's first grandchild – and Jim had disgraced himself by his choice of the new woman in his life. Their disdain for him was a powerful unifier.

'I don't think Jim could relate to a mature woman,' said Susan as they sipped liqueurs. 'That's why he always goes for twenty-year-olds. I don't think he's ever going to grow up. You'd think that by now he'd want somebody more . . . more . . . *adult*. She was his secretary, for God's sake – what a bloody cliché. It would be laughable if it weren't so sad.'

Thelma laughed anyway, swaying slightly. Her cheeks were flushed and her eyes bloodshot. These were precisely the thoughts she'd had about Susan a decade or so ago. 'I know – he's pathetic,' Thelma said. 'Still, as long as he carries on supporting your children . . .'

'He's got no choice, has he, thanks to child tax. Anyway, he can afford it.'

'True. Just as well, really. Especially if his latest little gold-digger manages to get herself pregnant.' Thelma hesitated. She'd called Susan a gold-digger at one stage.

'Anyway,' Thelma continued hurriedly, 'there was something else I wanted to talk to you about.'

'What's that?'

'I want you to put some pressure on that son of mine for me. It's his father – Dave – his real father. He needs help, but he's too proud to ask for it. You know that when the paternity test showed Dave was Dan's father, Dave sold his share of his and Jim's business . . . obviously they couldn't go on working together. And, as you know, Jim eventually got full control of the company. Well, Dave went downhill for a while after that – and I feel a bit guilty about him. Child tax was more severe then than now and he was only just beginning to get back on his feet around the time you and Dan started getting together.'

Susan was nodding. She knew all this and was impatient for Thelma to get to the point.

'I'm going to ask him,' Thelma said, meaning Dan, 'to try and find room for Dave in his company.'

'Dan and Dad,' Susan mused. She knew Dan wouldn't like the idea. Even though Dave's money had raised him, she knew he still thought of Jim as his father. If it hadn't been for the wedge she herself had driven between them, she knew Dan would prefer to link his business with Jim's rather than take on a poverty-stricken partner like Dave, even if he did share his genes.

'I'll have a word with him,' Susan said. 'See what I can do. After all, Dave is this little thing's grandfather.' She patted her stomach.

'Isn't it weird?' said Thelma. 'If it weren't for paternity tests, we wouldn't know that. I was convinced that Jim was Dan's father, you know. I still can't believe that one evening with Dave was all it took. Without those tests, we'd have been sitting here thinking that Jim was your baby's grandfather.'

'My God, I know,' said Susan. 'That's really creepy. I would have thought that Jim was *father* of my first two children and *grand-father* of my third.' She grimaced. 'You know, it's really difficult to imagine what it used to be like, not always knowing for sure who was your child's father.'

Thelma agreed. 'Not only that, think what it was like not knowing that the father, whoever he was, would *have* to help you support your baby. Just imagine trying to raise a baby on only your own money. How on earth did women manage?' She lifted her glass and thrust it towards Susan. 'To the man who invented paternity tests,' she said.

Susan chinked glasses with her. 'Except I bet it was a woman,' she said.

'Naaagh,' said Thelma confidently. 'Child tax – *that* was woman's work. Paternity tests – that will have been a man. I guarantee.'

2

The Weakening Bond

SCENE 2

Green Light to Infidelity

'Are you sure it was a red light this morning?' the man asked his partner as, sitting on his bed, he tied up his shoelaces. 'You should have let me see it, just to be sure.'

The woman didn't answer immediately. She was peering intently into the mirror, concentrating on applying her lipstick. 'I *can* tell the difference between green and red, you know,' she said eventually. 'I'm not colour-blind.'

Stepping away from the mirror, she turned to face him. 'There, how's that?'

He looked her up and down. 'Fairly incredible for someone who's had two children,' he said admiringly.

'Never mind the two children. How do I look, really?'

'Absolutely fantastic.'

He moved towards her as he spoke, intending to kiss her – but she backed away.

'Don't. You'll smudge me.'

A few minutes later, the couple went downstairs. He phoned for a taxi while she looked in on the baby-sitter who was refereeing a noisy but nebulous game between the two children.

'Try and get them to bed by eight,' said the mother. 'Nine at the latest.'

Minutes later, the taxi arrived, pulling into the drive of their large house. The driver sat with his engine running while the man opened the cab door for his partner. Before getting in, she kissed him gently on the lips.

'Wish me luck,' she said nervously.

'Of course,' he said, squeezing her hand affectionately. 'You can do it.'

'Get lots of work done,' she said. 'I'll see you in the morning sometime.'

The man stood in the house doorway and returned his partner's wave as the taxi backed out of the drive. Part of him wanted to run after her and tell her to stay – but the other more rational part knew she must go. They so needed her evening to be a success.

After closing the front door, he stood in the large hallway. He knew he should go straight upstairs to his study and begin writing. There was tremendous pressure on him to get his next book written quickly – and to make it good. His partner's eldest child, a girl now eight years old, was not his but the product of a casual relationship while at university. Her second child, a son now aged six, *was* his – and the pride of his life. Two hugely successful books in succession had bought them a large house in a good area and had seduced him into committing his son to a private education. In the interests of family unity, he also topped up his stepdaughter's child support so that she could go to the same school. Her genetic father didn't really approve – either of the principle of private education or of its cost – and paid the assessed level of child tax and no more. And as he was registered for three other children and earned relatively little, that didn't amount to much.

A mediocre book followed by an outright flop now seriously threatened the lifestyle he and his 'family' had enjoyed for the last five years. His main problem was that his enthusiasm for writing had gone – and the spark that reviewers had seen in his first two books had gone with it. Now, he had to force himself to write – and it was a painful process.

He took one purposeful step towards the stairs and his study, then hesitated and eventually turned on his heels to go into the kitchen and make himself a coffee. Hearing him, the two children ran in to see him, followed closely by the baby-sitter. It was 7.30 in the evening, and the children were already dressed in their pyjamas.

It seemed a luxury hiring a baby-sitter so that he could work on evenings that his partner went out to her party-political meetings, but they'd decided it was worth it. He could write 2000 words during the course of an uninterrupted evening – and on the basis of the money per word earned by his first two books, that would pay for the baby-sitter a hundred times over. And when the pressure was on, as now, there was no question of not having one.

'Go and start working, if you like,' said the baby-sitter. 'I'll make your coffee and bring it up to you.'

It was a bad evening. He sat staring at his computer screen, seeking inspiration. He wrote 500 words and deleted 1000. At 20.30, he heard the children being put to bed, took a five-minute break to say goodnight, and requested another cup of coffee. At 21.15, he heard the baby-sitter go into the children's room to quell a riot – and at 21.30 the house fell quiet.

Still he stared at the screen, trying to force himself to concentrate, but it was no good. His mind was elsewhere – about five miles away, he guessed – wondering how his partner was getting on with her quest for insurance. The meeting would soon be over, and everything would begin. Would it be the same hotel as the last few times?

By 22.00 he had 200 words to show for the evening – just about enough to pay the baby-sitter if this book did as badly as the last. He was wondering whether to quit when the door opened. The baby-sitter walked in carrying not coffee this time but a glass of wine for herself and a bottle of malt whisky for him. She was used to his evening routine.

After pouring his drink, then handing him his glass, she stood behind him so as to read the words on the screen. After sipping her own drink, she put it down on the table. Placing her hands on his shoulders, she kissed the top of his head and asked how the evening had been.

'Hopeless,' he said. 'I don't know what's happened to me.'

'Why don't you stop? Let's go to bed. You need your mind taking off it.'

'There's no rush,' he said. 'We've got all night.'

'But I want you. I've been sitting downstairs, thinking about you.' She nuzzled the back of his neck, then whispered in his ear, 'I'm all wet.'

He laughed and kissed her right hand. 'Go and have your shower. Let me finish this paragraph. I've been struggling with it all evening and I'm nearly there.'

She knew better than to argue. 'OK, but if you haven't finished by the time I've had my shower, I'll ravish you in the middle of your paragraph, ready or not.'

He watched her go, then immediately switched off his computer. He had no intention of working further – it was just that he wasn't in the mood for sex. His partner would just about be sitting down to her meal with the local Member of Parliament. Expensive food, expensive wine and a luxury hotel suite – at least, that's what it had been the first time – and the three times since.

Downing his whisky in one kamikaze swig, he picked up the bottle and walked into the guest bedroom. He and his partner had an unwritten rule that they never had sex with anybody else in 'their' bed. After undressing, he stood in front of the window in the dark room, watching the rain falling in the light of the street-lamps.

The baby-sitter came into the room, still drying herself. Then she stood by his side, the pair of them staring out of the window in silence, occasionally sipping their drinks.

'Is she with someone else?' she asked eventually, putting her arm round his naked waist and resting her hand on his hip.

He nodded – but couldn't tell her the full story. He couldn't admit that his partner was trying to conceive to a rich politician. Handing her the bottle, he said he was going to have a shower too.

'Warm the bed for me,' he said as he went out of the door.

Ten minutes later, he slipped under the bedclothes alongside her. She was sitting up, still drinking, and for a while he did the same.

'You still love her, don't you?' she said. It was more of a statement than a question, and she didn't wait for an answer. 'I know you do – but try and forget her, at least for tonight. Try and love *me* – just a little bit. You must feel something for me.'

He smiled, put his arm round her shoulder, and kissed her forehead. They both put down their glasses and held each other, enjoying the closeness. This was the third time they'd spent the night together, and it would be about the tenth time they'd had sex.

As far as the young woman was concerned, the writer was everything she was looking for. Rich – she thought – and famous; she couldn't believe her luck when the agency first sent her to baby-sit for him.

For a while, she enjoyed the closeness of just lying in his arms. But what she really wanted was action – and her strategically placed hand told her that action was still some way off. Impatient, she slid down the bed and began the massage that had never failed to work in the past – and didn't fail now. Five minutes later, he was moaning so loudly that she worried she'd gone on too long. Quickly, she sat astride him and before he could react, placed him inside her.

She thought her mouth had taken him to the point of no return – that he was so on the brink that he would need only seconds inside her to ejaculate. She thought his passion had taken over – that he couldn't stop now even if he wanted to. But she was wrong. The second he realized he was inside her his passion subsided.

'What are you doing?' he asked accusingly.

'What do you think?'

His arm flailed to the side, trying to open the top drawer of the bedside cabinet.

She grabbed at his arm, pressing her pelvis firmly down so that he stayed inside. 'I just want to feel *you* inside me for once, not eight inches of rubber.'

'No,' he said. 'We can't.'

'It's all right,' she said. 'It's quite safe. I had a green light this morning.'

She sensed him relax a little.

'Look, if you're worried,' she continued, 'just let me know when you're going to come and I'll let you out ... please. It's so nice having you inside me properly, for once. Skin against skin.' She moved a little, to emphasize her point.

He hesitated. It was the first time he'd been inside her without a condom and he was panicking. She began moving, trying to find the rhythm that would send him headlong towards a climax. He fought for self-awareness – to know when to stop – but failed miserably. When the moment came, he couldn't help himself. He actually thrust even harder and exhaled even more noisily. Secretly triumphant, she collapsed on top of him.

In the middle of the night, he was woken by yells from his son, but when he eventually dragged himself to investigate the boy was still asleep, crying in a nightmare. When he returned to bed, the baby-sitter pulled him into her arms and before he knew it they were being intimate once more. His half-hearted question, 'Are you sure it was a green light?' was his only pretence at contraception.

Before daybreak, she left his bed and dressed. After she'd gone, he slept for another hour before waking the children, giving them breakfast, and driving them to school. He had just sat at his computer in pretence at writing when he heard his partner open the front door. Joining her in the kitchen, he made them both a coffee, eager for news.

'It wasn't easy,' she said, 'but I've got them. They're in here,' she rested her hand on her stomach, 'swimming away. All we can do now is cross our fingers and hope it really was a red-light day.'

He smiled. He didn't like the thought of somebody else's sperm inside her – but they'd agreed it was necessary. If his next book – assuming he ever managed to finish it – was as big a failure as the last, they were in real trouble. They would have to sell the house and return their children to State education. But if last night's plan had worked, they could be receiving substantial child support from the politician for the next twenty years. It would give them the insurance they needed – just in case his writing career was a long time in resurrection.

'Maybe you should spend the day in bed. Take it easy. Put your feet up. Give his sperm a chance.'

'I will,' she said. Then, after a pause, she said, 'Stress! Stress is the key. Just don't let me get stressed.'

'Of course. Anyway, come on – tell me what happened. I assume he did take you to the hotel after the meeting, etcetera, etcetera, just like last time.'

She nodded. 'The only difficult bit was stopping him using a condom. He was as paranoid as we said he'd be. I tried everything. I got him really worked up, put him inside me to show him what he was missing – everything. But he wouldn't do it. He said he was too worried about getting me pregnant to be able to enjoy it.'

'So what did you do? Did you doctor his condoms, like we said?'

She shook her head and laughed.

'I didn't need to. In the end it was the silliest of things. After all the writhing around and getting him really excited – it was really silly. I nearly didn't bother – it seemed so stupid.'

'What?'

'Well, first I told him I'd had a green light in the morning – and the idiot believed me. Why will men believe anything that's vaguely technological? Then,' she laughed, not noticing her partner's wan smile, 'I promised I'd let him out just before he came. Of course, he left it too late and I was slow. I'm pretty sure I got the lot.'

She laughed for a while, then realized he wasn't joining in quite as much as she'd expected, given the success of her evening. Eventually she looked at him quizzically and said, 'Presumably the guestroom got used last night?'

He hesitated, then said, 'Yes, of course.'

'Just as long as you were sensible,' she added, unconcerned. 'Anyway, look, if I'm going to bed, you need to go shopping.'

Decline and Fall

Two urgent females and a couple of coy men; fertility predictor kits with red and green lights; a couple who condone infidelity – is this the end of traditional relationships between men and women? Is the nuclear family dead? At a guess, this scene, like the scene before it, is set around 2035 – and the answers are 'Yes' and 'Yes'.

Towards the end of the twentieth century, traditional family life in the Western world was changing rapidly. Fewer couples were marrying, the divorce rate was rising and the number of lone-parent and blended families was soaring. Many contemporary observers saw these changes as social and moral decay – but something that could be halted if only the subversive cause could be eliminated. The campaign against absent fathers, for example, had the unmistakable aura of a moral crusade.

Such a crusade is misplaced and futile, because far from being a sign of social decay, the demise of the nuclear family is an inevitable step in social evolution; humankind's innate response to the modern environment. Measures that many hope might oppose the 'decay' will actually hasten it, because they weaken still further the misunderstood biological forces that for a few centuries in a few places had made the nuclear family such a viable institution. Even as innocent a technology as a fertility predictor kit has the potential to weaken the nuclear bonds in the early twenty-first century. Relationships between men and women are about to enter a new era and will never be the same again.

Mate Guarding – Evolution's Nuclear Bond

To the romantic, men and women spend time together, put their arms around each other, hold hands, sleep together and live together because they love each other. To the cynic, they do these things to stake a claim on each other and to advertise to the world their exclusive rights. To the biologist, they do these things to make it difficult for their partner to have sex with anybody else. The behaviour is *mate guarding* – and it works.

The nuclear family is no more than the product of a man and woman – particularly the man – doing their best to prevent each other from being unfaithful. And as we saw in Chapter 1, the way to do this is to spend as much time in each other's company as possible. If the man succeeds, the children will be his, and he will be inclined to help raise them. If he fails, or thinks he's failed, he is more likely to become a deadbeat dad.

Biologically, the drive to hinder a partner's infidelity stems from self-interest and self-preservation. An unfaithful partner might bring home a sexually transmitted disease, picked up from his or her lover. Or, the partner might be lured away by the paramour, leaving the betrayed either alone or a lone parent. On balance, men have always had the bigger price to pay for a partner's infidelity. This is because in the past men with an unfaithful partner always risked a lifetime of unknowingly raising another man's child.

There is nothing modern or even uniquely human in this. These are ancient and universal penalties for being cuckolded; penalties as demonstrable for other monogamous animals as for us. It is not surprising, then, that both men and women have an evolved psyche that is biased towards opposing a partner's infidelity. The emotions are deep-rooted and cannot easily be masked by the conscious mind. If there is a risk of suffering, the pre-programmed psyche will surface and destroy even the most determined of 'open' relationships. Only when a partner's infidelity no longer threatens a person's interests does the jealous psyche relax.

As men face the greater costs from a partner's infidelity, they have been shaped to be more jealous, more possessive, and more aggressive in their possessiveness than women. Nevertheless, women still manage to be unfaithful occasionally – and from time to time children are born as a result. And they get their chance because no partner can possibly be vigilant all the time.

At first sight, being vigilant seems more difficult *for women* than for men. A watchful woman has the problem that her male partner is continuously fertile, continuously potent, and continuously interested in the women around him. All she can do is be as alert as possible for signs that harmless interest is metamorphosing into dangerous action. In contrast, a watchful man should find damage limitation much easier. True, his partner can have sex – and contract diseases or be lured away – at any time, at any stage of her menstrual cycle. But she is *fertile* for only a few days in each cycle. Surely, then, all a man needs to do is guard his partner intensively during her few fertile days and relax a little at other times. That way he could at least avert the danger of raising another man's child.

Unfortunately for men, evolution has favoured womankind over the matter. Natural selection has provided women with a very powerful weapon in their unconscious quest to collect the best genes and sperm around. They hide their fertility. The result is that men need to be just as attentive as women and just as much of the time. In fact, they need to be more attentive because they have more to lose if they fail.

If women did not hide their fertility, the nuclear family might never have formed in the first place.

Infidelity's Accomplices – Sexual Crypsis and Fertility Predictors

Sexual Crypsis

Some female primates – chimpanzees, for example – advertise when in their menstrual cycle they are most fertile. The skin around their anus and vagina swells and changes colour on those days when conception is most likely. In addition, they show much greater interest in sexual activity. This is the period of oestrus, or being 'on heat'. In such primates, dominant males show maximum interest and the most intensive mate guarding during the height of the female's oestrus. At other times they relax and allow the female to have sex with other males.

In contrast, women – in common with marmosets, orang-utans and various other primates – hide their fertile phase, a phenomenon known as sexual crypsis. In these species, with no indication of when the female is most fertile, a male has little choice but to try to guard her throughout her cycle. Forced to spread his attentiveness more thinly, he is inevitably less vigilant on those critical few days in each cycle when his mate really is fertile. This liberates her, endowing her with more freedom to be unfaithful at times when she can conceive.

Among primates, there is a clear link between the formation of a nuclear family and sexual crypsis. Most monogamous species *are* sexually cryptic. Moreover, sexual crypsis evolves first and monogamy then follows, rather than vice versa. It seems that a female primate is only prepared to allow one male to try to monopolize her if she has sexual crypsis to give her that extra ability to be unfaithful. Alternatively, a male is only prepared to spend all his time with just one female if he cannot tell precisely when she is fertile. Either way, without sexual crypsis there is no nuclear family.

Yet at first sight, sexual crypsis seems a fragile phenomenon. Surely it should be easy for a man – or at any rate the woman herself – to predict her own fertility. Why did the people in the scene need technological help to tell them when the women were fertile? It should be simple. As most people will remember from school biology lessons, a woman's menstrual cycle is around twenty-eight days long. On days one to five, she menstruates. On days six to thirteen, she prepares hormonally to produce an egg – ovulate – which she does on day fourteen. If the egg is not fertilized by a sperm, it dies on day fifteen or sixteen, and from days seventeen to twenty-eight, her body prepares for its next menstruation. We can add to this simple arithmetic the less well-known fact that sperm stay fertile inside a woman for up to five days. So we can conclude that the fertile phase of a woman's menstrual cycle should be the six- or seven-day period from day nine onwards – QED. Who needs technology?

So if a woman's menstrual cycle is so clear-cut why doesn't a man know precisely the days on which to keep his partner firmly within his sights? And where is the need for the fertility predictors portrayed in the scene? The answer to both is that in reality a woman's cycle is very different from how it is portrayed in school textbooks, and it is far from clear-cut. The female body has been shaped by natural selection to disguise when it is fertile – and as usual, natural selection has performed its task with great efficiency. Sexual crypsis is real and very efficient, and even modern technology finds it difficult to crack the code.

Fertility Predictors

Sexual crypsis was evidently a powerful but subconscious weapon in the hands of our female ancestors, endowing them with greater freedom to mate with whom they wished on their most fertile days. But for the modern woman with her careful, conscious and often strict timetable for family planning, sexual crypsis is a major inconvenience. When the modern woman decides it is time to conceive what she needs is to be able to conceive *now*, not in a year's time. Yet her body, chemically and behaviourally rooted in some ancient environment, is still working to an ancestral timetable. It is still waiting for the right man and the right moment to commit itself to parenthood. It still hides its monthly cycle, and it does not yield to modern needs at all easily.

In recent years, science has tried to come to the aid of the modern woman. In various ways it has attempted to strip away the cover-up engineered by millions of years of evolution. The technological aim is to tell a woman categorically two things. These are, 'Ovulation is ages away so you can have unprotected sex now with impunity,' and, 'You are about to ovulate, so have unprotected sex now if you want to conceive, avoid unprotected sex if you don't.'

The most recent systems for women to monitor their cycles

are computerized and in effect are sophisticated developments of the ovulation predictor kits that preceded them. The pioneering system *Persona*, launched in Britain in October 1996, uses simple urine tests along with other data such as the woman's average cycle length. By 1998, *Persona* was also marketed in Italy, Ireland the Netherlands and Germany.

On the *Persona* system, the woman needs to carry out a urine test eight times a month and to consult the monitor daily. The sophisticated software in the hand-held plastic device tracks two hormones and purports to tell a woman when she is fertile and when she is not. On the fertile days – signalled by a red light – she should abstain or use another contraceptive method unless, like the women in the scene, she actually wants to conceive. On infertile days, the green light gives the go-ahead for unprotected sex.

In developing fertility monitors, the two main aims for scientists were first, to predict the day of ovulation several days in advance, and second, to recognize when ovulation had occurred and the egg was dead.

The second aim was fairly modest. The post-ovulatory phase is reliably characterized by a heightened basal body temperature and a low-oestrogen, high-progesterone hormone profile. Not only that, but the time interval from ovulation to the first day of the next period is a predictable fourteen days. From the moment a technological system can confirm the woman has entered this phase, therefore – say three to four days after ovulation – there are ten or eleven consecutive *infertile* days before the next menstruation. *Persona*-type systems can then give a reasonably confident green light until the next menstruation.

The difficult bit, technologically, is advance prediction of ovulation. A woman's fertile phase stretches from a few days before ovulation to the day of – and maybe the day after – ovulation itself. How long before ovulation the fertile phase begins is determined by the fertile life of sperm once inside a woman. Earlier, for convenience, we took this sperm life to be five days. In fact, scientists cannot actually agree over this. Some claim that sperm are only fertile for two days, but there seem to be reliably documented instances of women conceiving five days after their last intercourse – and the most extensive study to date also suggested that the fertile phase lasts from five days before to the day after ovulation. On the principle that it is better to err on the side of caution when dealing with the risk of unplanned pregnancies, the task facing technology is to recognize the point at which ovulation is just five days in the future.

Throughout the pre-ovulatory phase, gonadotrophin-releasing hormone (GnRH) is produced by the hypothalamus in the brain. It stimulates the pituitary gland to release luteinizing hormone (LH) and follicle-stimulating hormone (FSH), which in turn stimulate the ovaries to produce oestrogens and release a mature egg. The levels of these different hormones can be monitored in blood and urine. *Two* days before ovulation there is a sudden surge of oestrogen, followed quickly by surges in LH and FSH. Any *Persona*-type system that detects these surges can confidently predict that ovulation is about two days away and would be justified in giving a sequence of three or four red-light days. And so far, that is about as far as the technology has got.

The major problem is that no change in a woman's body has yet been identified which can reliably predict ovulation from as far in advance as *five* days. Sex on any day during the pre-ovulatory phase of the cycle – even during menstruation – *could* be followed by ovulation within five sperm-life days and so *could* lead to conception. All that *Persona*-type systems can do to cope with this problem at present is to make probability assumptions based on the day of the cycle. Ovulation is less likely to be only five days away during, say, menstruation than by day nine. It is still risky, but not unreasonable, therefore, to give a green light for the first few days of the cycle but a red light from about day seven onwards – until ovulation is confirmed and the egg is dead.

There are two particular problems that make life very difficult for such technology. One problem is that women do not ovulate in every cycle. When ovulation occurs – so-called ovulatory cycles – the cycle is fertile. When it does not – so-called anovulatory cycles – the cycle is infertile. So far, there is no way of predicting in advance whether an upcoming cycle is going to be infertile – green light all the way – or fertile. The other problem is that there is a real possibility of women sometimes showing reflex ovulation.

Some research suggests that, once menstruation has finished, the woman's body in effect goes 'on hold'. Her body is poised to ovulate but is waiting to see what happens sexually. The trauma of rape or the excitement of intercourse with a new partner, a lover, or even a long-term partner who is home only briefly – such as a soldier on weekend leave – might all actually trigger ovulation *within the next two days*. It is even possible that anovulatory cycles are potentially fertile but the woman is never triggered to ovulate by sexual events. This 'on-hold' process – exquisitely shaped by natural selection – is probably what accounts for the notorious variation in menstrual cycle lengths (from fourteen to forty or more days) that occurs not only from woman to woman but also from cycle to cycle for the same woman. It is the pre-ovulation phase that varies in length, not the post-ovulation phase. It has been known for a long time that women's cycles are more regular during phases of sexual activity. The 'on hold' process is more likely to be broken if a woman has sexual contact with a man.

More basic research is still needed before technology can cope with all the vagaries natural selection has programmed into the menstrual cycle. Even high-tech systems such as *Persona* are still unreliable. As aids to *conception* they probably increase chances rather than decrease them. But as aids to *contraception* they are flawed. So much so, that in January 1998, the British government issued a health warning to women not to use *Persona* if an unplanned pregnancy was completely unacceptable. It warned that each year one in seventeen women who use the system could become pregnant because the device might wrongly identify a day as infertile – green light when it should be red. This failure rate compares with one in 100 women on the pill and one in fifty couples using condoms.

In the scene, though, the assumption was made that by 2035 medical research had finally cracked the crypsis that natural selection has imposed on the female body for millennia; the fertile phase could be accurately identified. The assumption was also made that a *Persona*-type system had been developed to translate the tortuous findings of medical research into a system – such as green and red lights – that is much easier for a confused public to understand.

If biotechnology does produce an efficient system for predicting fertility, though, what would this mean for the nuclear family, given the importance of sexual crypsis in its formation in the first place? The answer is that it won't help – but then nothing is likely to help. The nuclear family is about to suffer a take-over.

Lone Parenthood: the Take-over

The nuclear family is a biological institution, not some pawn of religion or politics. And like most biological institutions it will form under certain conditions and disintegrate under others – whatever moralists and legislators might wish.

For a nuclear family to form and be stable a woman has to gain from a man's help in raising her children and a man has to be unable to tell when a woman is fertile. Given both these factors, fear of the partner being unfaithful then becomes biological cement that binds the couple together. Women are vulnerable to unsupported conceptions and to being left destitute and men are vulnerable to raising another man's child. The solution for both is to spend as much time in each other's company as possible. Change any of these factors – by either of the two sexes becoming less vulnerable or even by the simple introduction of fertility predictor kits – and the biological bonds weaken. Beyond a certain point, they disappear altogether, and that is where the modern environment is leading.

The last few decades of the twentieth century saw the emergence of an institution destined to challenge the nuclear family and change the face of society. Lone-parent families metamorphosed from rarity to major force. Of the ten industrialized nations with recent statistics, Italy is changing most slowly, with only one in twenty-three children in homes without a father. Galloping away in front is the United States with one in five of its children living in lone-mother families. Of those born since 1980 in the United States, as many as half of white children, and eight out of ten black, will spend some part of their childhood in a lone-parent family. Britain isn't far behind and everywhere the transformation is gathering momentum.

If this trend continues into the twenty-first century eventually the majority of people will be living in lone-parent families. As over 90 per cent of lone parents are currently female, lone-mother – not nuclear – families seem destined to become the building blocks of future societies. Lone-father families will also become more common than now. For the moment, though, we can concentrate on the lone-mother scenario.

Most people today have a very negative attitude towards lone-mother families, and any suggestion that these are destined to become the social norm might seem like an indictment of the developments responsible. True, lone parenthood does currently have undesirable consequences *on average* for the survival, health and fertility of mother and children. There are also undesirable social consequences, such as children with poorer school performances and higher rates of delinquency. But just because something has worked sub-optimally in the past does not mean that it will continue to do so in the future. We need to know why lone parenthood has faltered in the past, and what might change in the future. We also need to know why men and women's ancient urges are suddenly generating lone-parent families in ever-increasing numbers. Are those urges so misdirected that they are pushing people into self-destructive situations? Or does lone parenthood have potential advantages in the modern and future industrial environments that will eventually become obvious to all? The answer seems to be that pessimism for the future is unnecessary. Lone parenthood *will* become the best system for raising children in the twenty-first century.

Of course, the change hasn't happened yet. For the moment, child-raising is still more successful in nuclear and extended families, with mother, father and maybe grandparents all helping with day-to-day care, than in lone-parent families. If that weren't the case, the nuclear family would be unlikely to have been such a popular institution in so many societies.

The past value of having both parents showing parental care is well documented. Studies of cultures as different as the Ache forest-dwellers in Paraguay in the 1980s and the inhabitants of Ostfriesland, Germany, in the eighteenth century have shown that the loss of a father is detrimental to a child's survival. The father's presence becomes increasingly important once the youngster passes the age of two. With a man to help, a woman's children are more likely to escape accident and disease and to grow into healthy, fertile adults, although at all ages the loss of a father is less damaging than the loss of a mother.

Even though it is clear that in the past a man has made a significant difference to a woman's success as a parent, we should not exaggerate the male role in parenthood. As fathers, men by no means rival the males of some other primates. In fact, from a primate perspective, humans are classed only as 'affiliators' not 'intensive caretakers'. Of the eighty human cultures included in a world survey of paternal relationships, fathers were rarely or never near their infants in 20 per cent and in only 4 per cent was there a close father–infant relationship. Even in these latter societies, such as the !Kung San of Southwest Africa, fathers spent only 14 per cent of their time interacting with their children. This is about the same as the three hours per day recorded for many fathers in industrial societies. However, some fathers in modern industrial societies spend as little as forty-five minutes *each week* interacting with their children.

There is nothing inevitably superior about biparental care – the raising of children by *two* parents – and the nuclear family it generates. In the wider biological perspective, parental care itself is rare and when it does occur *lone parenthood is the rule*, not the exception. In fish, frogs and toads the lone parent is usually the father. In birds and mammals, it is usually the mother. There are actually very few situations in which biparental care is effective. It just so happens that the human situation is one – or at least it used to be. For most animals, it would actually be counter-productive if both parents stayed to help. In mammals, in which the mother is the primary parent, not only can males rarely help but they also eat too much food and spend too much time squabbling. Fathers can be a day-to-day liability rather than an asset and once female mammals have conceived many are better off on their own.

We can even see hints of this for women in modern industrial societies. Much of the apparent disadvantage suffered by the children of lone mothers separated from a previous partner is measurable *even before* parental separation, and does not necessarily worsen afterwards. It is the quality of the family relationship, of which separation is only a part, which seems to be the major influence. It would seem, then, that some fathers are actual liabilities as live-in partners, and in such cases women can raise their children just as well or better on their own.

Why then do we find that in current society lone mothers struggle to match nuclear families in their parental performance? The answer is obvious, and gives a clear pointer as to why a future society in which lone parenthood is the norm should not be viewed with dismay. Analyses have shown that 80 per cent of the difference in parental performance is due to the difference in income between lone parents and nuclear families. The remainder is probably due to the lack of affordable day-care facilities. As long as lone parents are financially comfortable and have sufficient help with hands-on care, their situation seems to create a perfectly good environment for the raising of children.

The most telling statistic is that widows – women who are lone parents through the death of a partner and for whom proper financial support has usually been arranged – show none of the negative facets of lone parenthood that bedevil women who have separated. The survival, health and fertility of the children of widows are not impaired compared with those in nuclear families. Nor do their school performances and rates of delinquency differ. The clear implication is that it is not lone motherhood *per se* that causes the problems, but the pervading circumstances, especially finance. In particular, how much the family income declines after separation can be more critical than the income level itself. This is one reason why *widowed* lone-parent families, which suffer less if any reduction in income when the father 'leaves', are much less likely to suffer from their change in status than deserted lone-parent families.

Even if lone-mother but financially viable families can maintain health and performance in the absence of a man, surely the children suffer from the lack of a male role model? Evidently not

– and biologically it wouldn't even be expected. In primates, nuclear families are the exception rather than the rule. Instead, the young are raised by a group of females, cooperating to various degrees in the reciprocal raising of children. Often these females are related – sisters, aunts, nieces, grandmothers – but not always. The same is even true of many human societies, the children being raised by a collection of females in an extended family similar to that for many other primates. Both in humans and these other primates, males float in and out of the female groups, offering occasional 'paternal' care but most often simply going about their business of collecting food and trying to inseminate females. There is no reason, therefore, for children in industrial societies suddenly to have been burdened by natural selection with the need for a *live-in* male role model. There are plenty of males in the wider environment from which children can learn what they need.

So we shouldn't expect the absence of a live-in male role model to influence a child's performance in later life – nor does it. In modern industrial societies, children perform no better socially or psychologically if they are raised by both parents than if they are raised by a widowed mother (as opposed to a separated mother) or by a mother and grandmother. Nor does *how long* a child is without a live-in adult male make any difference.

So far we have concentrated on lone motherhood. How do men and women compare as lone parents?

At first sight, men seem to have the edge. Actual measures show that on average children are healthier when raised by lone fathers than with lone mothers (though this is no more than we should expect because on average lone fathers have a higher income than lone mothers). However, they only do equally as well in terms of social development (when again, on average, they should do better, for the same reason). On the negative side, daughters of lone fathers are more likely to conceive in their teens and the lone fathers themselves seem to have a higher mortality rate. When allowance for their higher income and greater ability to afford day-care is made, though, men in the United States and Europe emerge as marginally *less* competent lone parents than women. Elsewhere – studies in Ghana, for example – show that resources under the control of women are more likely to be devoted to children than are resources in the hands of men. Similarly in Brazil, income in the hands of a mother improved a child's health almost twenty times more than income controlled by a father – and also increased the child's survival prospects.

In late twentieth-century society, there are two rather separate groups of lone parents. One group is of women who are the victims of rape, of men 'having sex and running' or who have separated from their partner. This is the group currently most at risk to poverty and deprivation and most in need of child support. Given child support and day-care facilities, though, these women will raise their children more successfully without their child's father than with him. The other group is of women of independent means – often career women – who never had any intention of having a live-in partner. They chose lone parenthood, are quite happy with their situation, and can afford to raise their children without disadvantage.

At the moment the former, more vulnerable, group is in the majority. The expectation, though, is that over the next few decades the second group will increase considerably in force and number. This is because the main factor that changed in the second half of the twentieth century was the level of women's dependence on men. At the close of the century there was still a little way to go. The continuing struggle of lone parenthood for women and the dilemma of paternal uncertainty for men ensured that biology still favoured the nuclear family more often than not. But all seems poised to change, as Scene 2 has illustrated. The marriage of child taxation and paternity testing stamps out historic penalties. Women no longer need to suffer financially from sex and men no longer need to suffer paternal uncertainty.

Which returns us to the irony mentioned at the beginning of this discussion. The crusade for child support enforcement had an undeniable moral edge to it. There was a punitive element to the legislation, aimed at making men who left their families impoverished suffer. In all the polemic, there was a clear expectation that child support enforcement would make men less likely to desert their families and thus strengthen the bonds that kept the nuclear family together.

In the hands of ancient urges, however, child support and paternity testing have almost exactly the opposite effect. Together, they hasten the demise of the nuclear family, rather than slowing it down. In the process, they will lead to relationships between men and women that are quite different from those of the last millennium – and the millennia before that. And as a result, they will play a major part in shaping a quite different society, based on lone parenthood and relationships of convenience.

No longer will a man need to spend extended periods of time with a woman simply to guard against cuckoldry. Paternity testing will tell him what he needs to know. No longer will a woman need to suffer an inept, perhaps violent and increasingly undesirable man simply to avoid becoming destitute. Child support legislation will see to that. Couples will have little practical reason to stay together once the early spark of excitement has disappeared. As soon as the emotional and practical problems of trying to live happily together neutralize the initial chemistry of attraction, they can separate with relative impunity.

Today, the lone-parent lobby is still a minority voice – but it is growing fast and once paternity testing and child support enforcement begin to bite it will clearly grow even faster. There will be a threshold that, once crossed, will herald very rapid change. This is the point at which governments realize that far from disappearing, lone parents are not only here to stay but are also destined to be the bulk of the electorate. Then, forced by the sheer volume of the demand, governments and/or employers will have no choice but to provide *real* and economic day-care facilities for children. From that point, lone-parent families – and the blended families that arise when lone-parent families cohabit – will quickly become the social norm.

Relationships – a New Era

Along with changes in family structure, the early decades of the twenty-first century are likely to see a major change in the sexual politics of men and women, because ancient vulnerabilities over unwanted conceptions will also change. In fact, for those women who want it, a new era can begin. This era heralds new reproductive possibilities – or rather, old but risky strategies become imbued with new potential.

In every generation, there have been women who have tried ensnaring men of wealth and status by conceiving to them. In past generations, though, this was a dangerous game. Men of influence could easily deny paternity; an example of paternal uncertainty for once working to a man's advantage. In the future, however, the balance of costs and benefits must change. Paternity can be established with certainty and men can be forced to pay for any child they sire. Suddenly, womankind's ancient gamble can become a secure strategy – as both the women in the scene realized.

For maybe the first time in human evolution, an unwanted conception will become more costly for the man than for the woman. The future onus will be on the man to be coy and cautious, not the woman. A sought-after man in particular, unless he is happy for his child tax to rocket, will need to insist on contraception, leaving his pursuers no choice but to attempt to seduce him into unprotected sex. The ancient balance sheet of costs and benefits will have been turned on its head.

Although the costs of his actions will be obvious to a man's logical mind, his conscious carefulness will have to wrestle with the impetuous urges that evolution has given his body. When it comes to the sexual act, a man will still be innately the more urgent of the two sexes. How many men, sexually stimulated to the verge of ejaculation as in the scene, will be able to resist and withdraw while an attractive woman is urging him to climax inside her?

In any quest to conceive to a wealthy man in order to bolster flagging finances, a woman of the future will find fertility prediction kits invaluable. No longer will she need to raise suspicions – or risk boredom – by seducing her target time after time in the hope that she might eventually strike lucky in the lottery of conception. Instead, she can wait for the red light, then set out on her campaign, knowing that novelty and the element of surprise are on her side.

We shall embellish this thesis about the disappearance of the nuclear family in later chapters. There is no point in doing so now because the phase illustrated in the scene – in which seduction and fertility prediction are major reproductive activities – is actually likely to be short-lived. There are factors still to be considered that will render such behaviour unnecessary while making the end-result – a society of lone and blended families – even more certain.

These new factors first saw light of day in the 1980s, masquerading as treatments for infertility – as Part 2 will describe.

Part Two

An End to Infertility

3

Women, IVF and Surrogacy

SCENE 3

Ursula's Choice

Across the room, two drunken men dropped their trousers, bent down and treated everybody behind to the sight of their buttocks.

'They don't improve, do they?' Ilsa said to Ursula. They were at a class reunion, but considering they had been such close friends at school, they hadn't met that often in the fifteen years since.

Ilsa, always the brighter of the two, had gone on to university and carved out a successful career in marketing. A procession of brief and fruitless relationships had eventually led to her living for the past eight years with Michael, one of the directors of her company. Despite unprotected sex from almost the moment they met, five years went by with no sign of children. Tests eventually showed a malformed womb and no possibility of ever carrying a baby. Ilsa had been telling Ursula her story when the sight of hairy buttocks had distracted her.

'That was three years ago,' she continued, once the distant trousers had been pulled up and modesty restored. 'I still can't come to terms with it. I have spells where, night after night, I dream that I'm nursing a baby. And it's so real. I can actually feel it sucking my nipple. I can feel the wetness all over me. When I wake up and the baby's not there, I scrabble around looking for it. And when I realize there is no baby – that it's just a dream – I burst into tears. Night after night – I'm really not sure how much longer Michael can stand it. Honestly, Ursula, you don't know how lucky you are, having four kids.'

After leaving school, Ursula had become a secretary, settled down with a policeman, and devoted herself to raising a family. Three years ago, though, Fred had moved out. 'You wouldn't think you were so lucky,' she said, 'if you had four kids all fighting over who's going to sleep next to you. The other night, I let two of them sleep with me because they were crying and wouldn't stop, and within fifteen minutes of each other they threw up. I'd just cleaned up from one and changed the bedding when the second one followed suit. It went all over my hair, the pillows – everywhere.'

They both laughed.

'Well,' carried on Ursula. 'It didn't seem funny at the time. What I'd give to go to bed with a man – or two – instead of any permutation of four kids.'

'I bet there are men queuing up,' said Ilsa. 'They always used to, even when we were at school – and you're still really good-looking. I can't believe you've had four children.'

Ursula was pleased by the compliment, which she knew was no less than the truth. 'I was just lucky,' she said. 'I find pregnancy really easy – and giving birth. Even breast-feeding. Some of my friends struggled from the word go. Some of them felt really terrible for nine months then had horrible things happen to their body when they gave birth. Not me – I'd quite happily have another four. It's the looking after them afterwards that I'm not so keen on.'

She paused, wondering how much more to say, but as Ilsa didn't say anything, she carried on. 'As for men queuing up, that's what started my problems in the first place – but they're not so keen now that I've got four kids. They're quite happy to have sex, but nobody's offered to be my live-in partner.'

'But surely you're getting child support for them all?'

Ursula gave an odd flick of her head, shrugged and looked away for a second. A group of men, in a circle with arms around each other's shoulders, were beginning a schooldays football chant. At the centre of the circle were about twenty full glasses of beer.

'Of sorts,' said Ursula. 'Unfortunately Jimmy – that's the youngest – turned out not to be Fred's. That's partly why he left me. Well, that and the fact that I didn't know the surname of the guy who must have been the father. Not his proper surname, anyway. Only the name he told me. When the child support lot eventually found him on their database he turned out to be a con artist between jail sentences. He's got no money and ten other children scattered around the place. All I get is State money for Jimmy – and you know how pathetic that is. Fred's contribution isn't much better. He's not earning much anyway – and to make it worse, he's just had another with his new woman. So *his* support for each is nearly down to the minimum. Anyway, the upshot is that things are rather tough at the moment. I can manage, but I've got to be careful.'

Ilsa saw some irony in the way that they each had an excess of what the other wanted but couldn't have. She and Michael had plenty of money.

The idea occurred to Ilsa almost immediately, but she had little chance to think about it further. A brawl began as ancient playground rivalries surfaced between two of the men. It spread quickly. Ursula and Ilsa watched in fascination for a while until a beer glass smashed against the wall just inches from them, at which point they left.

Back home, and excited by her thought, Ilsa actually woke Michael to talk to him. They had toyed with the idea of surrogacy ever since realizing it was the only path open to them. They had ruled out adoption from the beginning – they both wanted their own child, their genetic child. And as the doctor had explained to her, with her womb in the state it was, surrogacy was the only solution. She just needed to borrow another woman's womb for nine months.

The only thing that had held them back was finding a surrogate they could trust. They had heard so many stories of surrogates from agencies changing their mind at the last moment and keeping the baby instead of handing it over. It made them hesitant to commit themselves. But to Ilsa, Ursula seemed perfect – and when she explained the situation to Michael, he too was enthusiastic.

They invited Ursula and her children to stay for the weekend. When Michael met her, he wasn't quite as convinced of her trustworthiness as Ilsa. But like most men, he responded to her warmly in other ways. The thought of having her carry his baby for nine months quite appealed to him. Eventually, he and Ilsa made Ursula an offer she would have found difficult to refuse under any circumstances.

'That's incredibly generous,' she said. 'I can't believe you would pay me all that money just to carry a baby for you – *and* add little Jimmy to your child tax as well. Of course I'll do it. I'm thrilled. It'll help you and it really should solve all my money problems. It's fantastic. When do you want to do it?'

The three saw a great deal of each other over the next few months in between visiting clinics, being screened for innumerable diseases, having medical tests, seeing counsellors – and solicitors. They all became close friends and often visited the clinic together. Ursula opted for a natural cycle so at least was spared the worst of the sniffing and injecting of hormones that Ilsa had to suffer in the run-up to having her eggs harvested.

'Are you excited?' Ursula asked Michael during a snatched bar lunch.

They were on their way back from one of their last visits to the clinic. It was two days before Michael and Ilsa's IVF – and according to Ursula's screening it wasn't many more days to her ovulating. Ilsa had visited the clinic with them but was feeling sick – so

Michael had taken her straight home afterwards. He'd then offered to give Ursula a lift home as well before going back to work.

'Of course I am,' he said. 'I've waited a long time for this. Anyway,' he continued flippantly, 'I'm looking forward to going in that room again to do my bit. They've got some great videos in there.'

Ursula laughed. 'Come on,' she said. 'I bet you don't really need to watch dirty videos to manage it.'

Michael laughed with her. 'Well, maybe not – but it all adds to the sense of occasion. Anyway, it's not that easy. I bet if I shoved you in a room and told you that you couldn't leave until you'd made yourself come – and that if you failed everything might fall apart – I bet you'd find it a bit difficult.'

She leaned towards him and whispered wickedly, 'Actually, I probably wouldn't have any trouble at all. It's one thing I'm rather good at.'

The pair laughed a laugh of closeness.

When the day came, Michael had no problem performing. There was no need to resort to his frozen sperm. Ilsa produced seven eggs, all of which were successfully fertilized *in vitro* and the embryos were placed in the deep-freeze. Ursula ovulated earlier than expected – on the day after IVF, in fact – and three days later had three of Michael and Ilsa's embryos placed in her womb. The other four were left in cold storage in case they were needed in the future.

A few weeks later, an ultra-sound scan confirmed that Ursula was carrying triplets. There was a fresh round of consulting and counselling and the trio eventually agreed to a selective abortion, leaving Ursula to carry twins. She wasn't entirely happy with the prospect of carrying twins, but accepted, not least because they renegotiated the contract so that she received twice as much money for her lump sum and 50 per cent more support for little Jimmy, her 'fatherless' child.

More tests followed. Cells were removed from the amniotic fluid around each baby as doctors checked for genetic and developmental abnormalities before deciding which of the three embryos to abort. Eventually Ursula, Ilsa and Michael were called back to the clinic – together.

This time they were uneasy. They were even uneasier when they arrived to find a counsellor in the room as well.

'We seem to have struck a slight problem,' said the ageing specialist once the pleasantries were over. 'Now, don't worry, all three embryos are fine. They're all perfectly healthy. A bit crowded together, perhaps, but otherwise . . . fine.'

'So why are we here?' asked Michael anxiously.

'Well . . . I'm not quite sure how to put this . . .' The doctor paused, gave a meaningful look at the counsellor who looked down at the notes on her lap, and then looked back, not at Michael but at Ursula. 'We seem to have found a peculiarity in the DNA pattern.'

'But I thought you said they were all right,' said Ilsa, feeling excluded – and panicking.

'Yes, they are. That's quite right. When I say DNA pattern, I mean DNA *fingerprints*. It would seem . . .' he paused yet again. 'Well, it would seem that . . . That . . . well, that not all the embryos have the same parents.'

Michael's mind worked with the speed of light and produced an instant anger. 'You mean you've messed up, don't you? That's what you mean. I've heard of this happening before. You didn't use my sperm, did you? You got my sample mixed up with somebody else's. Bloody hell.'

The doctor raised both hands and patted the air. 'Calm down,' he said. 'Please, calm down. No, that's not what happened. In fact, there's no question about it. All the embryos are yours. You're the father all right.'

'Oh my God,' said Ilsa. 'You got the eggs mixed up. You lost my eggs and used somebody else's. The babies aren't mine. After all of this, they're not mine.'

'Please,' said the doctor. 'Please – let me finish.'

'Well, for Christ's sake get on with it, man,' said an exasperated Michael. 'What are you talking about?'

'How can I put it?' mused the doctor. 'All of them are yours,' he said, looking at Michael. 'Two of them are yours,' he added, looking at Ilsa. 'And one of them . . . is yours.' As he finished, he looked at Ursula, peering at her over the top of his spectacles.

There was silence as they each wrestled with their thoughts. Ursula and Michael stole a brief, telepathic glance at each other. But it was Ilsa who eventually broke the silence.

'I'm sorry,' she said. 'But I really don't understand how that's possible.'

'To be honest,' said the doctor, 'neither do I – though there are several, I'm sure remote, possibilities that we can discuss later if you wish. But by far the most urgent matter now is, what do we do next? If we are to go ahead as we all agreed and reduce the numbers . . . well, which one – or two – do we remove? Normally, we would remove the smallest, or the most awkwardly placed. But if we do either, then it means that you . . .' he looked at Ilsa ' . . . will have only one baby. In fact, the strongest and best-placed embryo . . .' he looked at Ursula ' . . . appears to be yours. So – what shall we do?'

Every Woman Fertile

In vitro fertilization and a surrogacy arrangement that goes wrong; if it weren't for the futuristic child support arrangements, the scene could easily be set in the twentieth century, not the twenty-first. The reproductive technologies described already exist and are gaining momentum, despite public apprehension. Even the reimbursement arrangements would sit quite comfortably within the modern-day United States – though not Britain or France, for example. On balance, though, we should probably place the scene around 2035, contemporary with that in Chapter 2.

Assisted-conception technologies such as IVF and surrogacy were developed to rid humankind of a scourge that is as old as reproduction itself – infertility. This they will do – and very soon. Few people could reprimand the Ilsas, Michaels and Ursulas of this world for the collaboration just described. In the same situation, most people would see the logic and appeal of surrogacy – even if they never followed it through.

Part 2 is divided into four chapters. First, we explore the way that current technology heralds an end to infertility for women. Then we do the same for men. The third chapter deals with one of the most bizarre of our potential reproductive options, the transplantation of testes. And the final chapter deals with what in many ways is the ultimate solution to infertility – cloning.

Fertility and Infertility: the Legacies

Fertility

Going from the outside inwards, a woman's reproductive system consists of one vulva (vaginal opening, labia, clitoris), one vagina, one womb, two oviducts (often called Fallopian tubes) and two ovaries. All the bits are joined except for the ovaries. These are detached from the rest, each being suspended a short distance from the opening of the nearest oviduct. This system has been handed down, relatively unchanged, from the earliest monkeys, spanning 50 million years of female ancestors. Basically natural selection has shaped the vagina to receive a penis and collect sperm, the womb to transport sperm and hold a foetus, the oviducts to make it easy for egg and sperm to meet – and the ovaries to produce eggs in the first place. And more often than not, the process works. Most women – nearly 90 per cent – are fertile.

Not that fertile women all conceive the moment they have unprotected sex, of course. Only one in three fertile couples manage conception in the very first month of a campaign and, on average, a healthy, fertile couple will take four or five months to conceive. Such averages, though, hide a wealth of perfectly normal and adaptive variation.

For a start, the chances of conception vary with age. A woman becomes pregnant most easily at the age of eighteen or nineteen, with little real change until the mid-twenties. There is then a slow decline to age thirty-five, a sharper decline to age forty-five and a very rapid decline as the woman nears menopause. Below the age of twenty-five, around 90 per cent of women conceive within six months of unprotected intercourse whereas above the age of thirty-five, less than 20 per cent do so. This lifetime pattern in part reflects how many eggs a woman produces each year at different ages. When about twenty years old, she is producing about five eggs per year. Her rate climbs to about nine by the time she is thirty, but by the time she is forty it's back down to about five and by fifty it is virtually zero.

At any age, even a fertile woman can have temporary phases when she finds conception extra difficult. Poor nutrition, temporary obesity, chronic fatigue or high fever can all reduce fertility for a while. One of the most important factors is stress. Whatever its source – financial hardship, crowded living conditions, accident, family illness or death, a partner's infidelity – stress is a powerful contraceptive and just about every aspect of fertility can be affected. Miscarriage, failure to implant, failure to ovulate, loss of libido – all become more likely if a woman is stressed.

For such women, fertility returns as soon as conditions improve and stress is relieved. But some women suffer from a problem that is as old as fertility – permanent infertility.

Permanent Infertility

Infertility is usually defined as the inability to deliver, carry or – most often – conceive a healthy child. In the scene, Ilsa's main problem was that her womb would not allow her to go through pregnancy. Depending on her precise problem, it might also have made it difficult for her to conceive. In fact, in her case, nearly every part of the normal road to fertility might have been impaired.

A worldwide study of human infertility carried out by the World Health Organization and published in 1990 concluded that about 15 per cent of humans are infertile. In industrial countries the figure is nearer 10 per cent and is roughly equal for men and women. The result is that about one in six couples find it impossible, or at least extremely difficult, to conceive. Out of every 100 cases of infertility, about forty can be traced to problems in the female, forty to problems in the male, and the remainder to conditions in each partner that interact to cause sterility.

Infertility can result from problems at any stage of the repro-ductive process, as we shall discuss in detail shortly. Of all the different forms of female infertility, blocked oviducts are the most frequent. A study coordinated by the World Health Organization in twenty-five countries and involving more than 10,000 infertile couples suggested that such blockages were a problem in over one-third of female infertility cases. Failure to ovulate came a close second.

It might seem surprising that such a high level of infertility exists given that for millennia natural selection has favoured only the most fertile of women. Why hasn't evolution rid the human population of such a fundamental problem as infertility? Part of the reason, of course, is that many of the factors that cause infertility are very recent in origin. But there is also a biological reason – because the main cause of infertility is itself an ancient legacy.

Infertility and Disease

Over half the cases of human infertility are due neither to a failure of the body, nor to modern artefacts. They are due to disease – infections of the urinogenital system that find their way into oviducts, causing inflammation and blockages and hence infertility.

Sometimes such infections cause only temporary bouts of infertility, the normal reproduction process returning once the infection subsides. Others, though, can cause enduring blockages and hence permanent infertility. Such diseases include what might appear to be the most innocuous of infections during childhood – passing irritations that are scarcely noticed. But if a female is unlucky, a single bout of such a passing urinogenital infection early in life can cause a lifetime of infertility. Other diseases, though, can cause more than infertility, and are transmitted sexually.

All animals suffer from sexually transmitted diseases (STDs). One of humankind's most unpleasant legacies from its biological past is a menagerie of organisms that have evolved to cash in on the most intimate of moments. Those concerned are a mixture of viruses, bacteria, chlamydiae, mycoplasmas, fungi, protozoans and even insects.

Records of some of the STDs that infect modern humans are as old as literature. There are ancient references, attributed in the Old Testament to Moses, to the 'Botch of Egypt' (probably syphilis) – and 2400 years ago Hippocrates described 'strangury' (gonorrhoea). The first records of other STDs are more recent. The most notable of these is AIDS (acquired immune deficiency syndrome), the first record for which is of a British sailor who died in Manchester, England, in 1959.

Some STD organisms are relatively benign, such as crab lice, but others are much more serious. These range from the single-cell types that cause non-specific urethritis, the bacteria that cause gonorrhoea and syphilis, to the viruses that cause genital warts, cervical cancer and AIDS. Beyond a particular stage of infection many of these diseases can lead to permanent infertility – not to mention death – particularly if they are left untreated.

Infertility on the Rise

Reports indicate that infertility rates in industrialized countries have been rising for the past three decades. From 1988 to 1995 alone, the number of American women of childbearing age who suffered from fertility problems jumped from 4.9 million to 6.1 million, a 25 per cent increase.

Part of this increase may actually be an artefact. It could be caused by an increase in the number of couples seeking medical assistance rather than an increase in the number who are infertile. It could also be caused by women increasingly delaying childbirth until their thirties, hence experiencing age-related problems we have already mentioned and will discuss in more detail in Chapter 7.

In part, though, the increase may be real. For example, it

could be due to an increase in the use of birth-control pills and intra-uterine devices for contraception. The latter in particular can cause infertility – both temporary and permanent. It could also be due to an increase in STDs, although this hasn't been established.

The Drive to Reproduce

Given an opportunity to reproduce, *most* infertile women – like Ilsa in the scene – will jump at the chance. It would have taken a hard person to deny Ilsa – or the millions of women like her – the opportunity to have her own genetic baby.

Without medical assistance, a woman who has not conceived after two years of unprotected sex has only a one-in-four chance of conceiving. At present, the cause of a couple's problem can usually be identified – about 90 per cent of the time – and can often be corrected – about 50 per cent of the time. The success rate, though, is forever increasing and before long every form of infertility will be treatable.

Producing children is one of the most basic of instincts, pro-grammed into the body and brain by 400,000 million years of evolution. Humans have inherited their urge to reproduce from the earliest self-replicating life forms in the primeval soup. Each one of us is the current representative of our own unique, per-sonal lineage that stretches back over countless generations. And every single individual in our direct lineage reproduced – other-wise we wouldn't be here. Little wonder, then, that from the moment of our conception our genes grind out instructions for an anatomy, chemistry and behaviour that make us nothing less – and some would say nothing more – than reproduction machines. And sooner or later, most of us do end up reproducing, perpetuating our personal lineage for one more generation.

It is no surprise to find that people, whether they say they want to have children or not, nevertheless show behaviour that almost inexorably leads them to have them. Nor should we be surprised to find that people, when threatened with the possibility of *not* reproducing, clutch at every straw modern technology can offer to help them. That is the way evolution has shaped the human psyche – and the more people feel that they are being barred by circumstance rather than their own disposition, the stronger the urge seems to become. Each new method is gratefully embraced as it becomes available. But while the infertile minority sanction each new technique by usage, the fertile majority manage to shake their heads and foresee problems at every turn.

In the sections that follow, we shall consider the three most widely discussed forms of treatment: fertility hormones, IVF and surrogacy. Each one was developed to treat a particular form of female infertility, and each one has triggered its own branch of public suspicion.

Failure to Ovulate, Fertility Drugs and Multiple Births

Failure to Ovulate

If there is no egg, then there is no baby.

Most women have a million eggs in their ovaries at birth and 250,000 by puberty. This should be plenty, because the maximum a woman could ever release is about 500. As a rough guide, each ovary normally produces an egg about every two months – one ovary one cycle, the other ovary the next – though the pattern can be confused considerably by natural anovulatory cycles.

In advance of each ovulation, when everything is working properly, six to twelve primary follicles of the thousands present in the ovary begin to grow. After about a week of growth, one of the follicles begins to outgrow the remainder, which usually give up and later disintegrate. Just before ovulation, the surviving follicle grows to a diameter of about 10–15 millimetres. Ovulation itself is the moment when the wall of the ovary ruptures and the egg is expelled from the follicle.

As we saw in Chapter 2, the whole ovulation process is under the control of hormones – from the hypothalamus, pituitary and ovary itself. Problems with the production of these hormones, or disease in the ovaries, can all prevent ovulation. If the problem is hormonal, the answer is in principle straightforward – restore the appropriate hormone balance by administering drugs.

Fertility Drugs

A number of 'fertility' drugs are in use, designed to stimulate reluctant ovaries to ovulate. As we shall see later, they are also used as part of other infertility treatments, such as IVF.

Clomiphene citrate, for example, can be used to stimulate the ovaries to produce one – or more – mature eggs in each cycle. It is sometimes also prescribed for women with irregular periods. The drug's manufacturer says that its use is associated with a 6–8 per cent risk of a multiple pregnancy. Another ovarian stimulant in frequent use, human menopausal gonadotrophin, is said by its manufacturer to carry a 40 per cent risk of a twin pregnancy.

Beyond a certain point, the dose of such drugs necessary to trigger any given woman in any given menstrual cycle into producing just one egg per cycle is largely a matter of guesswork. And as doctors and patients both prefer to err on the side of some eggs rather than no eggs, an inevitable result is the risk of multiple births. And multiple births are on the increase.

Multiple Births

Women, like most primates, have been programmed by natural selection to carry and give birth to just one baby at a time. Some biological provision has been made for producing twins – a large enough womb and two breasts, for example – but little arrangement has been made for having triplets or above. The evolutionary logic is simple. With an active and mobile lifestyle, combined with slow development of infant mobility, primate babies, once born, have to be carried everywhere. Carrying one is difficult enough while foraging or running from danger. Carrying two or more is almost impossible and can have damaging and long-lasting effects on the health of any mother that tries to do so.

Even so, multiple births do sometimes occur naturally in all primates. In humans, most multiple births involve twins – about once in every eighty-nine births. By contrast, triplets naturally occur about once in every 7900 births and quadruplets about once in every 705,000 births. Until 1997, the most famous multiple births were the Dionne quintuplets, five identical girls born in Ontario in May 1934. The quints became global celebrities and three Hollywood movies were made of their lives. In the middle of the Depression, sales of Dionne dolls outstripped those of Shirley Temple. In 1985, septuplets were born in California but one was stillborn and three died within nineteen days. In 1993, healthy sextuplets were born in Indiana. Then, in Iowa in November 1997 came the climax – so far – of the story of multiple births. After taking a fertility drug, Bobbi McCaughey gave birth to four boys and three girls – the world's only live septuplets.

The first study to look at the problems surrounding multiple births was published in 1990. Many of its findings have fuelled a decade of public disquiet over the phenomenon. Records showed a considerable rise in the 1980s in the number of quads-plus births as a result of the use of fertility drugs and other reproductive technologies. Between 1982 and 1989 the number of such births more than doubled – from twelve sets to nearly thirty per 100,000 deliveries.

Not only were more triplets and quads being born in the 1980s than in the 1970s, more were also surviving – though not without considerable assistance. Half of the mothers of quads-plus had complications with their pregnancy – and almost all had at least one hospital stay before the birth. The babies tended to be premature and to suffer from their crowding in the womb. Half of quads-plus births occurred before thirty-two weeks of gestation, well over a month early, and the great majority of births – of triplets as well as of quads-plus – were by Caesarean section.

As a result of their prematurity and crowding in the womb, more than half of the quadruplets weighed less than 1500 grams at birth. Some died shortly after birth. Even of those that survived, over half spent a month or more in intensive care. Finally, sometimes years later, infants delivered at a multiple birth were found to have an increased risk of cerebral palsy.

A pregnant woman diagnosed as carrying triplets, quads or more has a number of difficult decisions to make. Should she terminate the pregnancy – or should she carry on, running the gauntlet of medical complications for herself and her babies? Or should she – like Ursula in the scene – wrestle with the idea of sacrificing some foetuses so that others can survive? In Britain, the process known as 'selective foeticide' or 'selective reduction' is supposedly available in only a few hospitals – and no public records exist to indicate how many are performed each year. The legal status of the process is uncertain, but at present it is covered best by the Abortion Act.

Any woman faced with a multiple birth and poor health prospects for all her hard-earned children would have difficulty deciding which foetus or foetuses to sacrifice so that one or two may survive and prosper. It would be especially difficult in the absence of any clear clinical reason for choosing one rather than the other. In the scene, Ursula, Ilsa and Michael were confronted with a virtually impossible task that could easily destroy their surrogacy agreement.

Ursula's choice was almost without solution, as various other situations involving surrogacy have proved to be. Before discussing surrogacy, however, it is more convenient to discuss IVF, which is an essential part of at least some forms of the surrogacy process.

IVF – a Helping Hand for Eggs

The commonest cause of infertility for a woman is some form of physical or chemical barrier in her reproductive tract – a barrier that either prevents sperm from reaching the egg or prevents the fertilized egg from reaching the womb. This was the suite of problems that IVF was initially developed to solve, and since 1978 it has been doing so with ever-increasing success.

In natural, unassisted, conception sperm are deposited in the vagina during intercourse, then swim through the cervix into the womb where they receive assistance in reaching an oviduct. They then have to swim up the oviduct to a zone two-thirds of the way along where fertilization can occur if they meet an egg travelling in the opposite direction. The fertilized egg then travels on down the oviduct into the womb where it implants and initiates pregnancy.

Almost every step in this sequence can go wrong, leading to infertility and the need for assisted reproduction technology, such as IVF.

Infertility Due to Sperm Never Reaching the Egg

Anything that prevents egg and sperm from meeting inevitably leads to infertility. There are several possible causes, the main ones being problems in the cervix, womb and particularly the oviducts.

When a couple have intercourse in the missionary position, the man deposits a pool of semen on what we can imagine is the floor of a chamber that forms at the far end of the vagina. The cervix – the neck of the womb – dips down from its normal position in the 'roof' of this chamber, and dangles in the seminal pool. There is a narrow channel running through the cervix through which sperm must pass if they are to get inside the woman's womb. This cervical channel, though, is not empty. It is filled with mucus that flows like a glacier slowly down through the cervix towards the vagina. Bit by bit, the bottom, older end of the mucus glacier drips out of the cervix and into the vagina.

Cervical mucus is vital to a woman's health. Its main function is to allow sperm through while keeping bacteria, viruses and other disease organisms out – a balancing act it achieves partly by its acidity, partly by its consistency and partly by its rate of flow. Fast-swimming sperm can make headway against the glacier's flow and can get through but the slower-moving bacteria and viruses cannot and get flushed out. However, the balance is delicate. If the mucus is too acidic, too impenetrable or too fast-flowing, even sperm cannot get through and the woman will be infertile. If it is not acidic enough, too easily penetrated or too slow-moving, disease organisms can get through and again the woman can become infertile, as we have seen.

Even once through the cervix, the sperm have the difficult job of getting to the top of the womb and finding the tiny entrances to the two oviducts. They probably cannot do this unaided – or at least, only slowly and with difficulty. Normally, the woman gives them a free ride. In effect the sperm surfboard up the womb on the crest of tiny muscular ripples that transport them almost to the oviduct. Structural abnormalities in the womb or problems with the nerves and muscles that produce these ripples can prevent sperm from reaching the oviducts. Once again the woman is infertile.

Even if the sperm reach the oviduct, a blockage or break in the tube may prevent them from swimming up to the fertilization zone, in which case, they may still never have the chance to meet an egg. By and large, though, blocked and broken oviducts are likely to render a woman infertile not because of their effect on sperm but because of what they do to the egg.

Infertility Due to the Egg Never Reaching the Womb

An egg has a volume that is over 30,000 times greater than a sperm's, so incomplete blockages in the oviduct are more likely to halt the egg than they are sperm. Sperm may still get through to fertilize the egg but the fertilized egg is unable to complete its journey to the womb.

Normally, when an ovary 'pops' during ovulation, the expelled egg is slowly wafted across the short space between the ovary and the open mouth of the nearer of the two oviducts. It is carried on a current created in the woman's body fluid by tiny hairs on the inside of the oviduct. Like a waiting hand, the finger-like projections that surround the entrance to the oviduct funnel the egg into the tube. From here, the egg begins its journey down towards the womb, a distance of about 8 centimetres, still carried along on the current created by the same tiny hairs.

In a fertile woman, the egg's journey to the womb takes about five days. But if the oviduct is blocked or broken, the egg will never arrive. Occasionally, blocked or damaged oviducts can be cleared, mended or bypassed by surgery, but if this is not possible the woman will be infertile.

Or at least, she would have been – in the days before IVF.

IVF – the Technology

Biologically, there is nothing unnatural about eggs being fertilized outside the female's body. Many animals habitually do so. Most marine invertebrates – from ragworms to sea urchins, for example – simply liberate their eggs and sperm into the environment and leave them to fend for themselves.

Over a span of about 3700 million years, the lineage leading to humans also used to fertilize eggs outside of the female body – and now it can do so again. There was a relatively short period, though – from the first of our reptilian ancestors about 320 million years ago until the last quarter of the twentieth century – during which fertilization inside the female's body became the norm.

Of course, internal fertilization is still the norm. But there is already more than a hint that before the end of the twenty-first century the human lineage will have reverted almost wholesale to external fertilization, thanks to IVF.

IVF – *in vitro* (literally 'in glass') fertilization – is the fertilization of one or more eggs outside a female's body. The technique has been used extensively in animal embryological research for decades, but not until the 1970s was it successfully applied to human reproduction. Before then, the possibility of children being conceived outside their mother's body was only the stuff of fiction.

The British biologist J. B. S. Haldane, for example, foresaw the development of babies outside the womb and named the process ectogenesis. In his fantasy of the future, *Daedalus, or Science and the Future*, published in 1923, Haldane anticipated that the first ectogenetic child would be produced in 1951; he was premature by twenty-seven years. Aldous Huxley elaborated the idea in *Brave New World*, published in 1932, in which he gives a vivid description of the Central London Hatchery, where eggs and sperm are stored in test-tubes for *in vitro* fertilization.

In fact, the *scientific* foundations of modern IVF techniques were laid down as long ago as 1890 – long before either of these two books. It was then that Walter Heapes demonstrated that embryos could be successfully transferred from one rabbit to another. Fertilization of a mammalian egg outside of the mother's body, though, dates back only as far as the 1950s and experiments with rabbit eggs. Then, in 1959, external fertilization and embryo transfer were finally combined. The first successful birth from an embryo formed *in vitro* was again with a rabbit. Human pregnancies from IVF techniques were first reported in the 1970s. Then in 1978 the first test-tube baby – Louise Brown – was born in Oldham, UK.

Human IVF programmes often differ in methodology between clinics. Usually, though, women are maintained on a drug-induced ovulatory cycle while further drugs are used to promote the growth of multiple eggs, thus ensuring that many will be available for fertilization. When ripe, eggs are removed from their follicles on the ovary surface by rupturing the follicle with a needle and sucking the egg into the needle's lumen. Sonographic egg recovery uses ultra-sound guidance; laparoscopic egg recovery retrieves the eggs via a small incision in the abdomen.

Most often, while the woman is giving up her eggs, the man is in a specially designated room doing what is necessary to ejaculate a sample of sperm. Given the pressure of the occasion, many men find they need the assistance of pornographic magazines or videos. Once the sperm have been collected, they are washed of all seminal fluids and resuspended in a sterile medium designed to mimic the fluid inside the oviduct.

After the eggs are retrieved, they too are placed in a special fluid medium. Then sperm that have been washed and incubated are placed with them and left for approximately eighteen hours. Fertilization usually takes place in specially prepared culture media in Petri dishes, not test-tubes. The eggs are removed, passed into a special growth medium and then examined about forty hours later. If the eggs have been fertilized and have developed normally, the embryos are transferred to the woman's womb. Following embryo transfer, progesterone injections may be administered to the woman daily to encourage implantation.

Typically, multiple embryos are transferred to increase the chances that at least one will implant. However, even if ten or more eggs are harvested and fertilized, usually only two or three are placed in the womb, as they were in the scene. If more than four eggs develop into embryos, the donor may have the option of cryopreserving (freezing) the remaining embryos for thawing and replacement in a later IVF cycle.

As many as 80–85 per cent of patients may successfully have eggs fertilized *in vitro*. However, only about 25 per cent will achieve a clinical pregnancy and a number of these will have a spontaneous abortion. The probability of a viable pregnancy is approximately 20 per cent with one IVF cycle.

Various modifications of the basic IVF procedure have been tried in attempts to improve success rates or to treat increasingly specialized medical conditions. In gamete intrafallopian transfer (GIFT) the harvested eggs and sperm are placed directly into the oviducts – on the womb's side of any blockage – with fertilization occurring in the woman's body. In zygote intrafallopian transfer

(ZIFT), the procedure is similar to GIFT, but the beginning-stage embryos (zygotes) are placed directly in the oviducts.

One of the most recent developments is a technique that keeps embryos growing for a few extra days in the Petri dish. Previously, clinicians had to put *in vitro* embryos into the womb when they were just one or two days old and still relatively fragile. The deadline for transfer was fixed because after two days the embryos' metabolism changes, rendering standard growth mixtures useless for nourishing them. A new culture mixture from the United States keeps cells growing *in vitro* for up to five days, making it much easier to pick out the strongest embryos. The hope is that soon, instead of transferring three, four or five embryos back into the womb, it may be possible to use just one or two.

Another recent development is the successful storage of eggs. Sperm can be stored easily but eggs proved to be much more fragile and over the years, attempts to freeze and thaw them almost always ended in failure. In October 1997 in Georgia, USA, however, a patient gave birth to twin boys conceived from eggs – another woman's – that had been frozen for more than two years. Successes have also now been reported in Italy, Germany and Australia. The technique is still expensive and unreliable – only two births in the first twenty-three tries in Georgia – but will inevitably improve over the next few years.

IVF – the Problems

Naturally, such a revolutionary new technology as IVF and its various offshoots was not accepted without controversy, and there are still many people who worry about its implications. Here we concentrate on the medical worries that have been expressed and leave the more general, ethical questions until Chapter 15.

There has been some concern for the health of women who expose themselves to IVF procedures. Most are treated with gonadotrophins, powerful but unpredictable drugs that lead them to produce several eggs. Three per cent of women who take gonadotrophins will suffer from the potentially lethal condition of ovarian hyperstimulation; there is also a suspicion that the drugs may increase the chances of ovarian cancer. Any danger, though, seems to be marginal and well within the limits of what most infertile women are evidently prepared to risk for the chance of having their own baby.

Most medical concern, though, has been for *the children* conceived via IVF. In particular, it was feared that such children might be at a higher risk of developing abnormally. The main reason for such a fear was the lack of sperm selection in IVF. In normal reproduction, sperm have to surmount many hurdles within the female reproductive tract before having any chance of meeting an egg. We have already described some of these hurdles and shall discuss others when we consider male infertility. IVF, though, literally hands sperm an egg 'on a plate'. True, in separating sperm from seminal fluid in the first instance, some IVF techniques retain only those that have swum to the top of a column of liquid and survived centrifugation. But even this 'swim-up' technique is still nowhere near as challenging as the obstacle course sperm have to survive inside the female. IVF babies, therefore, might be the products of unfit sperm – an origin that might lead to increased risk of abnormal development.

Analysis of nearly 1600 births in Britain between 1978 and 1987 showed that, on average, IVF babies were significantly more likely to be premature and hence lighter than babies conventionally conceived. Largely, this was due to IVF often producing a multiple pregnancy – because in order to increase the chance of *any* pregnancy, more than one embryo needs to be transferred to the womb. There was no evidence, though, that the incidence of abnormalities was higher among IVF babies. In contrast, in

Australia the data suggest that IVF children may be two or three times more likely to suffer from certain birth defects, notably spina bifida and heart abnormalities. There is also a suggestion that some of the drugs used to stimulate ovulation may enhance the risk of birth defects as well as ovarian cancer. But the number of births recorded in both Britain and Australia is still too small for scientists to know whether any of these differences are real or simply the result of chance.

Ideally, worldwide data need to be put together, but funding is difficult to raise – national governments, the EC and the pharmaceutical industry have little interest. In the meantime, some comfort can be taken from the studies that have been completed. If it is so difficult to demonstrate a real difference between IVF and conventional babies, then any extra risk from IVF – if there is any at all – must be very small.

Public debate about IVF is likely to continue for decades to come, but it will be largely academic. It is far too late to deny the process a place in the future of human reproduction. IVF has been unequivocally sanctioned by usage. In the United States alone, more than 33,000 babies have already been born via IVF and related processes. In 1994, for example, there were 7000 births.

IVF is here to stay.

Surrogacy – and Bottle-feeding?

In the scene, Ilsa may or may not have been able to conceive. Whether her eggs were being fertilized by Michael's sperm month after month was unclear and is actually irrelevant. This is because

Ilsa was one of the bands of infertile women whose main problem lay in her womb – with either implantation or gestation. As a result, Ilsa could never have become a mother to her own genetic child without help. Her dreams would always have been of the baby that never was. But in her case, even IVF was not enough to solve her problem on its own. Surrogacy was also needed.

Infertility Due to Failure to Implant

Successful fertilization of the egg is a major step in normal reproduction, but it does not guarantee fertility. The fertilized egg might pass straight through the woman's womb without implanting.

Almost immediately egg and sperm have joined, the fertilized egg starts to divide as it heads on down the oviduct. By the time what was once a single-cell egg reaches the womb the new life consists of a 'blastocyst' of about 100 cells. The mother's body prepares for the arrival of this tiny bundle of cells by thickening the lining of her womb (from 1 millimetre to 3–4 millimetres) and by increasing its blood supply – steps that are all under hormonal control. If the egg has been fertilized, the blastocyst normally implants in the womb lining two to four days after arriving from the oviduct. While waiting to implant, the blastocyst receives nourishment from the mother via secretions from the womb lining – again under hormonal control. At the same time, the blastocyst develops special *trophoblast* cells on its surface. These cells secrete enzymes that digest and liquefy nearby cells in the womb wall. The trophoblasts then multiply rapidly, and invade, digest and imbibe further cells from the womb lining as they begin to form the foetal half of the placenta. This 'invasion' of the mother's womb lining is known as implantation.

If anything goes wrong between the blastocyst's first arrival and secure implantation there will be no pregnancy. The blastocyst either dies in the womb or simply passes straight through. Even fertile women do not always implant a fertilized egg but it is difficult to know how often they fail. It is thought that although most such eggs survive to reach the womb, only about 58 per cent on average begin to implant. Even then, some quickly fail and perhaps only 42 per cent implant securely enough to survive to the twelfth day of pregnancy. The actual proportion is sensitive to both levels of stress and the woman's age.

Some women, though, habitually fail to implant. Malformations of the womb or hormone imbalances that affect the womb lining can all lead to infertility at this stage of the process. In the scene, Ilsa may have had this problem, as well as the next described.

Infertility Due to Spontaneous Miscarriage

About one in five foetuses are miscarried during the first three months of pregnancy. To the distraught mother, such an event seems like a nightmare – yet there can be positive elements. This is because a large proportion of foetuses miscarried in this way have some form of developmental deformity.

Most of the *development* of a baby's organs takes place during the first three months of pregnancy. After that, gestation is largely a matter of the baby's growth. Any abnormality in the baby's development usually results, therefore, from an event in the first three months. The mother's body is actually quite good at detecting abnormalities in the developing child – and when it does, its reaction is often to abort the foetus spontaneously, preparing the ground to try again as soon as possible. And usually, the next attempt is successful.

Some women, though, habitually miscarry during the first three months, as if their defence mechanism persistently recognizes the developing foetus as being damaged in some way. Of course, in some cases, repeatedly abnormal foetuses may well be the cause. More often than not, though, the problem is again either a malformed womb or a hormone imbalance. Once the pregnancy is past the first three months, it usually has an excellent chance of producing a healthy baby. Miscarriages are then most often the result of accident or disease – or occasionally undue stress.

Pregnancy in humans lasts on average about 270 days (from conception to birth). Throughout this time, the baby is floating within amniotic fluid, held in a tough amniotic sac, and mother and baby are communicating chemically across the placenta. The baby is held in the womb by the strong fibrous walls of the cervix. Come the moment of birth, though, the cervical walls have to soften, stretch and weaken, the pubic bones have to widen, the amniotic sacs have to burst, and the baby has to be pushed through the cervix, between the bones, along the vagina and so out of the woman's body. After the birth the woman has to shed the placenta and rejuvenate the womb lining as well as return all her organs to their non-pregnancy state. More often than not, the whole process works smoothly – albeit painfully – and the mother is able to go through the entire process again on a future occasion. Occasionally, though, it does not and the woman is thereafter sterile.

Some women are born with a weak cervix or other problems with their womb, others develop such problems through accident, illness – or even childbirth. Whatever the cause, they habitually have problems completing the final stages of pregnancy. In the scene, Ilsa had such a problem. IVF alone, therefore, would not have helped her. No matter how many developing embryos were placed in her womb, none would have produced a baby. They would either have failed to implant or would have been miscarried at an early stage.

Ilsa was quite capable of producing fertile eggs and Michael was quite capable of producing fertile sperm. But to realize their ambition of becoming parents, they needed a womb that was capable of carrying a baby for nine months. In other words, they needed a surrogate.

Surrogacy – the Technology

Surrogacy is an arrangement in which a woman carries a child on the understanding that it will be handed over to another woman after its birth. In some cases, there is no technology involved at all; in others either artificial insemination or IVF is involved. This variation is because there are several different situations that can be helped by surrogacy. Each situation generates a different set of relationships between the various adults and the child produced following the surrogacy agreement.

In its most basic form, surrogacy is no different from adoption. An infertile couple reaches an agreement with a fertile woman – the surrogate. This surrogate agrees first, to conceive a child – by her (the surrogate's) own husband, partner, or other man of her choice – and second, when that child is born, to hand it over to the infertile couple to raise. Such an arrangement might be desirable if either or both the man and the woman in the commissioning couple cannot benefit from IVF. Examples would be a man without testes or a woman without ovaries. In this basic arrangement, neither member of the commissioning couple is the genetic parent of the child they raise. They are, in effect, adoptive parents.

Perhaps the most common use of surrogacy in recent years, however, has been for couples like Ilsa and Michael in the scene, where only the woman is infertile. Just how surrogacy is used – and its genetic consequences – depends on the nature of the woman's infertility. If she is unable to produce eggs then the usual arrangement is for the surrogate to conceive via an act of condoned infidelity by the commissioning man. Except that, to make the act seem less like infidelity, the commissioning man and the surrogate do not usually have intercourse. Fertilization is achieved instead by artificial insemination, as described in Chapter 4. Whether they have intercourse or use artificial insemination, the result is the same. The child that the commissioning couple eventually receives is the man's genetic offspring but not the woman's. He is the child's father, but she is, in effect, its stepmother.

The options are different if the commissioning woman can produce eggs but cannot gestate a baby because of problems with her womb. Now the commissioning couple can produce an embryo by IVF but instead of the embryo being placed in the womb of the mother, it is placed in the womb of the surrogate. In this situation, the child is the genetic offspring of both the man and the woman of the commissioning couple. Genetically, at least, they are both fully biological parents. This was the situation with Ilsa and Michael in the scene.

Private arrangements for surrogacy, without legal involvement, have probably been taking place for centuries, if not for millennia. In fact, every time throughout history that, via infidelity, a man has had a child by a woman other than his partner which he and his partner then raise as their own, the process is a form of surrogacy. The first case of a woman giving birth to a baby created from a donor egg via IVF, however, was in California in 1984.

Surrogacy in species other than humans goes back to Walter Heapes's pioneering experiments on embryo transfer in rabbits in 1890. The most recent development for other animals is to use surrogates who are a different species; a horse, say, to gestate a zebra. To ensure that such cross-species pregnancies take, researchers perform a bit of transplant surgery on embryos from both horse and zebra. They swap the outer trophoblast so that the inner, embryo-forming cells from the zebra are encased in a trophoblast from a horse. The horse surrogate mother then accepts the developing zebra embryo as if it were one of her own species. This embryonic surgery has great advantages for breeders of endangered species. One female of a rare species can produce many embryos at once, which domestic animals can then carry to term.

Nobody has yet suggested that humans should use other primates as surrogates – even though it might solve many of the problems we are about to discuss. Nor has anybody yet suggested that a woman should act as surrogate to some other species of primate. Theoretically, though, both are possible. What has been suggested is that humans should make their surrogacy arrangements not with another species of primate, but with a machine. So far, though, the only kids to have begun life in an artificial womb are to be found at Tokyo University. They are Japanese goats.

The first, relatively unsuccessful, attempt to create an artificial womb took place in France. In 1969, a sheep foetus was kept alive in one for two days. In 1992, Japanese scientists removed a goat foetus from its mother by Caesarean section after 120 days' gestation, about three-quarters of the way to its full term. It was then placed in a rubber womb filled with artificial amniotic fluid, and successfully 'delivered' seventeen days later.

The main aim of such research, of course, is not simply to produce an artificial surrogate. Artificial wombs will make it easier to study foetal physiology – but they will also allow sick foetuses to be saved that could not be saved by present treatments. One use for humans would be to help foetuses in the final stages of multiple pregnancies when the womb becomes too cramped – or to save a foetus that is too young to be treated by normal intensive care but whose mother has been killed in an accident.

At present, the Japanese researchers believe it may be possible to incubate a goat's foetus from as early as ninety days into gestation, and a human foetus from about the sixteenth week. It can only be a matter of time, though, before the whole of the gestation period can be accommodated in an artificial womb. In winter 1997/8, *Time* magazine predicted that the first human to be conceived via IVF and 'carried' to term in an artificial womb would be born – if that's the right word – in the year 2022. But for the time being, surrogates remain normal, red-blooded women. And therein lie most of the problems.

Surrogacy – the Problems

Of all existing methods of assisted reproduction, surrogacy attracts the widest opposition – though doubtless, before too long, this dubious accolade will be usurped by cloning. When things go wrong with surrogacy, everybody hears about it. The first two examples of surrogacy arrangements that went wrong – each a landmark on its side of the Atlantic – serve to illustrate the point.

The first case in the USA, that of 'Baby M' in New Jersey, in the mid-1980s, is perhaps the most famous, because it was the first time that a failed surrogacy really hit the headlines. A married biochemist, William Stern, signed a contract with the intention of having a child to raise to which he was genetically linked. In return for $10,000 in 'compensation for services and expenses', the surrogate, Mary Beth Whitehead, was impregnated with his sperm and subsequently delivered a three-day-old paternity-tested baby, her second daughter, to the Sterns. She then changed her mind. She wanted the child. However, after nearly two years of litigation and with their surrogacy contract voided as against the public interest, the Sterns obtained custody. Significantly, though, the New Jersey Supreme Court decided that biology should determine parentage. William Stern was pronounced Baby M's father and Mary Beth Whitehead her mother.

The first case in Europe was that of 'Baby Cotton', also in the mid-1980s. An American couple (husband fertile, wife infertile) arranged with an Englishwoman, Kim Cotton, for her to bear the husband's child. Nothing went wrong with the arrangement as such but the story caught the attention of the press. Because of the public outcry – largely whipped up by the media – the local authority attempted to remove the child to a place of safety. The child was made a ward of court, and a judge had to determine what would be in the baby's best interest, deciding eventually in favour of the commissioning parents. Legislation – the Surrogacy Arrangement Act – swiftly followed the Baby Cotton case making the operation of *commercial* surrogacy agencies in Britain a criminal offence. Kim Cotton, who was Britain's first surrogate mother, went on to form COTS (Childlessness Overcome Through Surrogacy) a *non-commercial* agency that relies to a large extent on public donations.

Cases such as these hit the headlines, create a public outcry and trigger demands for surrogacy to be banned – even to be made illegal. We need, though, to keep a sense of perspective. Such cases are rare. The majority of arrangements in which surrogates do the job required of them and the commissioning parents end up with the baby they long for rarely attract media attention.

An exception was the case in Britain of a former Roman Catholic nun. In May 1997, forty-year-old Theresa McLaughlin – by then a child-care worker – found herself on the front page of British newspapers when it became known that she had acted as surrogate mother to five babies, including twins. She became pregnant six times in as many years, miscarrying twice. Three of the children went to one couple.

Theresa McLaughlin had become a nun at the age of eighteen but left her order four years later. She married and had a son but her marriage ended after two years. She first became a surrogate mother in 1989, bearing a son for a Swedish couple who subsequently adopted him formally. Two years later she had a daughter for an English couple and in 1993 gave birth to a boy for a childless Australian man and wife. The following year, after two miscarriages, she had twins – a boy and a girl – for the same couple. Before being put in contact with the British surrogate, this Australian couple had reportedly spent £60,000 seeking treatment and surrogacy in America but without success.

Theresa McLaughlin's story illustrates a number of important points. First, all seven of the people involved in these surrogacy arrangements benefited. There were no losers – at least not biologically and probably not financially either. All Theresa McLaughlin's surrogate pregnancies were achieved via artificial insemination (a practice we shall discuss in connection with male infertility). This she performed herself using sperm donated by the would-be fathers. Biologically, therefore, hers is a clear success story. In effect, she had six children with four different men – children that were hers genetically as well as by birth. In the process, she recruited considerable assistance in raising those children, especially those for which she was a surrogate. Financially, it has probably been less successful. She is reported to have received only £15,000 in total over the years to cover expenses and loss of earnings, because current regulations in Britain outlaw anything other than out-of-pocket expenses for surrogacy. In the United States, surrogacy is much more of a commercial enterprise.

The commissioning *men* involved also clearly benefited biologically, producing children that would otherwise have been denied them, at least with their current partners. To reproduce, they would at the very least have needed to be unfaithful to their partners. The commissioning *women* may not have gained biologically from the arrangements – still being genetically childless – but should at least have gained psychologically. Moreover, their actions may have curbed their partners' temptation to be unfaithful or even to leave them.

COTS arranged all Theresa McLaughlin's pregnancies. In fact, she was the agency's first surrogate. By 1997, COTS had arranged 208 surrogate births, and although the former nun held their record for most births, several of the women on their books had had three babies. Although COTS have arranged surrogacies that went wrong and attracted media attention, the agency would claim that the vast majority of the couples and surrogates that have passed through their books have benefited, not suffered, from their existence. In the United States the equivalent organization, CSP (Centre for Surrogate Parenting, Inc.) had arranged 579 surrogate births by March 1998 of which 194 involved IVF.

Given that most of the people who enter into surrogacy agreements benefit, it is difficult to pinpoint precisely why the whole process generates so much hostility from people lucky enough not to need such arrangements. Most of the hostility is emotional rather than logical, but real enough for all that.

Historically, most cases of surrogacy have been either planned adoption or condoned infidelity. Is it really so much worse to arrange to adopt a child in advance of its conception rather than wait and see what is available?

As far as condoned infidelity is concerned, all five of the children for whom Theresa McLaughlin was both a surrogate mother and a natural mother were the result of just such behaviour, with artificial insemination taking the place of intercourse. Unlike most children born as a result of male infidelity, though, these five were raised by their fathers and stepmothers rather than by their mother. Such an arrangement can occur naturally in any number of ways without involving surrogacy and without generating any public outcry or demands for male infidelity to be made illegal.

Would it really have been more acceptable to the general public if the three commissioning men had been secretly unfaithful to their partners with Theresa McLaughlin, then abandoned her to raise the resulting child single-handedly? In the scene, the fact that Ursula conceived to Michael – we assume via intercourse – would even in 2035 probably have seemed far more underhand an act to Ilsa than would any surrogacy involving condoned infidelity via artificial insemination. It would certainly have seemed more underhand than the agreed process of IVF.

Of course, the forward planning involved in surrogacy arrangements imparts a cold-hearted element to the process. Emotionally, many people find such detachment difficult to accept. Yet how many of the people who feel this way about surrogacy feel the same way about marriage? But there is just as much forward planning associated with most traditional marriage ceremonies, which in principle at least are supposed to mark the beginning of sexual relations between a man and a woman. Does this mean that months-long forward planning is somehow less detached and less cold-hearted for sex than for family planning?

Public angst is always greatest when surrogacy arrangements fall apart. Sympathies and allegiances are pulled hither and thither as either the surrogate mother or the commissioning couple tries to renege on their agreement. It is almost as distressing a spectacle for the observers as it is for the trio of people in the thick of it.

The commissioning man, of course, has no major psychological conflict over surrogacy. The male psyche has been shaped to accept that fatherhood involves having a child that comes out of somebody else's body. Few psychological acrobatics are needed for a man to accept that it is *his* child coming out of a surrogate. In fact, as Michael betrayed in the scene, the prospect can be nearly as appealing as infidelity.

Conversely, it is not surprising that from time to time a surrogate mother decides she wishes to keep the child. She, too, is at the mercy of ancient urges. To the female psyche, any child to whom she gives birth is hers, without question. And throughout our evolutionary past until recently – in fact until 1984 – she will always have been correct. Even now, in most case of surrogacy, the surrogate *is* the natural mother of the child in every possible way. And every so often, her hormones and psyche together will simply not let her part with the child.

Things have been slightly different since 1984 – ever since surrogacy began to involve IVF so that the egg as well as the sperm could come from the commissioning couple. Biology now tells us that the baby has two mothers – a genetic mother and a birth mother. Psychologically, though, the birth mother may still feel that the baby is entirely hers, because ancient urges can often override modern logic. Equally, the genetic mother may have difficulty convincing *herself* that she really is the mother, because she did not carry and give birth to the child. Both women may struggle to make logic overturn the ancient legacies that are their psyches. It is not surprising that conflicts sometimes arise. Nevertheless, just because surrogacy arrangements sometimes break down and lead to an unseemly tug-of-war between the participants is no reason for a public demand that surrogacy be made illegal. Marriages also often break down and lead to an unseemly tug-of-war. In fact, they sometimes lead to violence and even murder. Yet nobody is demanding that marriage be made illegal.

So who has the greater right to the child – the surrogate or the commissioning couple? As with divorce settlements, surrogacy settlements simply require a fair and informed legal super-structure to decide who should have what and under what circumstances. In fact, the questions that have to be answered are more or less the same. What was the legality of any contract made? What is a fair solution to the conflict? What is in the child's best interests? These are the questions that really need answers. And one of the first is to decide an order of biological rights for the women concerned. What is the order of priority over claims to the child should a dispute arise? Should the genetic mother have prior claim? Or the birth mother, in cases of egg donation when genetic and birth mothers are not one and the same? Or is it the commissioning mother? The first two have biological claims, the last has a legal claim.

The same question has to be resolved for the men involved. Who has priority, the genetic father or the commissioning father? More often than not, this tends to be the same man. It is still necessary, though, to decide whether the genetic and/or birth mother has priority of claim over the genetic father. Now that paternity can be established with certainty (see Chapter 1), the age-old acceptance that the mother automatically has the greater claim perhaps needs to be re-addressed.

Surrogacy has the potential to help too many people to be banned simply because governments or lawyers cannot work out a fair way of dealing with the occasional problems that arise. There is an onus on governments and legal systems to stay apace with developments in infertility research, not simply to resist them because they raise inconvenient questions.

One of the most puzzling aspects of the surrogacy issue is the view held by the decision-makers in some countries that surrogate mothers should not receive payment for their services, other than, at most, modest expenses. The principle, it seems, is that surrogacy arrangements are acceptable as long as no one makes a profit from them. It is as if payment turns the surrogate into a prostitute. This strange principle is discussed in Chapter 13 alongside other financial implications of infertility treatment.

Bottle-feeding – the Relevance, the Paradox

Hostility towards the surrogate over payment is just one element in many people's emotional reaction to the phenomenon of surrogacy. Another element is hostility towards the commissioning woman for delegating part of her reproductive responsibility to another woman. It is almost as if pregnancy and labour are chores that every woman should perform before she deserves to become a mother. And being born infertile is no excuse.

Historically, this is not the first time that the delegation of a reproductive chore has been met with public hostility. Surrogacy is not the first example of women handing over part of their reproductive responsibility to other women and triggering public concern and outrage in the process. Wet-nursing, in which from antiquity to modern times a woman who cannot or does not want to breast-feed allows another woman to do it in her place, had a similar reception.

During the seventeenth century Puritan theologians devoted sermons and large tracts of popular conduct books to the evils of mothers who did not breast-feed. The analogy is not quite perfect. The mother who opted out of breast-feeding attracted more hostility than the wet-nurse who did the job for her, whereas in surrogacy arrangements the surrogate attracts more hostility than does the commissioning mother. Otherwise the two situations are quite comparable.

Despite public and theological hostility, many women opted out of breast-feeding during the eighteenth and nineteenth centuries. More would undoubtedly have done so had wet-nurses been more generally available. Then, during the nineteenth century, scientific discoveries about infant feeding followed by the commercial availability of baby milk formulas meant that the abandonment of breast-feeding snowballed.

In non-industrial cultures all women breast-feed their children for an average of 2.8 years (but up to five years in some cultures). In contrast, the majority of women in industrial societies nowadays avoid breast-feeding. In Britain in 1990, for example, although 60 per cent or so of new mothers made some attempt to breast-feed, within a fortnight of birth the figure was down to

50 per cent and after six weeks was down to 40 per cent. Only one in ten women breast-fed past nine months. Bottle-feeding has been sanctioned by usage – and in becoming the norm it has taken public opinion along with it.

Gone now is the sense of public and theological outrage of the seventeenth century. In fact, a few decades ago in Europe and the United States breast-feeding was actually portrayed as unhygienic and most women were dissuaded from natural nursing – by the medical profession as well as by their peers. Despite recent medical attempts to reverse the trend, there is no real sign that breast-feeding each child until the age of two or three will ever again become the norm.

On the contrary, in many places in the United States, nursing mothers encounter disgust, hostility and prudishness over breast-feeding. Nursing mothers in the United States have been asked to leave shopping malls, restaurants, public parks, courtrooms, and so on, where *bottle-feeding* of babies is regularly seen and accepted: they have even been threatened with arrest for indecent exposure. Public opposition to breast-feeding is so pervasive that several states have had to enact laws clarifying that breast-feeding is not a form of indecent exposure. Moreover, the claim that there is something sexual and improper about a woman breast-feeding her child past early infancy has caused some mothers to have reports of abuse filed against them for doing so – though nobody as yet has been found guilty.

If a woman cannot or does not want to breast-feed, she can delegate the job to another woman, the wet-nurse, or a piece of technology, the bottle. If a woman cannot or does not want to be pregnant and give birth, she can delegate the job to another woman, the surrogate, or in the future to a piece of technology, the artificial womb. We don't yet know whether surrogacy or the artificial wombs of the future are harmful to any of the people concerned. We *do know* that bottle-feeding is harmful to both mother *and* child. Yet surrogacy generates hostility and concern whereas bottle-feeding is not only sanctioned by usage, it is positively reinforced by public response. Will there ever come a time when the prospect of a woman gestating her own child similarly generates public disgust and abuse?

This intriguing contradiction between social attitudes over surrogacy on the one hand and bottle-feeding on the other is discussed further in Chapter 15.

IVF and Surrogacy – Another Marriage

Apart from the futuristic child support arrangements that placed Scene 3 around 2035, we can now see that it could have been set in any year from 1984 onwards. That was when the two main technological developments portrayed – IVF and surrogacy – were first used together. The only action that might be considered futuristic in some countries was the handsome payment that Ursula received for her services.

The likely impact of IVF and surrogacy on society and relationships is portrayed in several later scenes. The main concern of Part 2 is to describe developments in the treatment of infertility and to indicate the extent to which they might herald an end to infertility. The combination of IVF and surrogacy has the potential to solve the problem of infertility for 99 per cent of women. The remaining 1 per cent – women who are infertile because they were born without ovaries or with ovaries incapable of producing eggs – cannot yet be helped. Even at the end of the twentieth century, they were still condemned to childlessness or adoption. Just around the corner, though, are technologies that can help even these women as we shall describe in Chapters 5 and 6.

For women, then, an end to infertility is truly on its way – and as we shall now see in the next chapter, so it is for men.

4

Men, Artificial Insemination and IVF

SCENE 4

Pros and Cons

Friday: 14.00

The man gazed forlornly out of the window and wished it would open. The hottest day of the year so far and off had gone the air-conditioning; the sweat was already soaking into his shirt under his arms. He was just fantasizing about taking off his jacket and rolling up his sleeves when the door to his office opened and his secretary walked in. Despite her black tights, she still managed to look cool and trim in her short black wrap-around skirt and white top.

'The first couple are here, Doctor. Shall I show them in?'

She was a temp – only there for a week while his regular receptionist was away on holiday. She'd worked in the building on and off for other people before, but not for him. Now today was her last day – and he would miss her.

'Yes, OK,' he said with resignation in his voice, moving towards his chair and desk. He could think of a dozen things that he'd rather do on a hot summer's afternoon than talk to couples about their problems.

'Oh, Marilyn,' he called, just before she disappeared. 'Can you bring me a strong, black coffee? And I guess we'd better have a jug of iced water and three glasses.'

He sat down and prepared to look caring and professional, wishing he hadn't drunk beer at lunchtime. He was in severe danger of falling asleep in mid-consultation. Everything was so routine these days. There were no challenges any more.

Looking uncomfortable, the couple came slowly into his room and sat opposite him at his desk.

'I'm sorry about the temperature in here,' he said. 'The air-conditioning has chosen today of all days to pack up on us.'

As he spoke, Marilyn came into the room with a tray laden with coffee and iced water.

'Please take off your jacket if you like,' he said to the man as his receptionist left the room. 'Make yourself comfortable.'

As the couple settled themselves in their seats, he examined their file, stifling a yawn halfway through.

'So,' he said eventually. 'You actually wanted to do it all yourself – very good – but things just didn't work out, eh?'

The couple preened themselves a little, feeling some pride that they had at least tried.

'Well, it is a mystery,' he continued. 'As your own doctor will have told you, it's not at all clear why you haven't simply got pregnant naturally,' he said to the woman. 'You're young, you're producing eggs regularly, your hormone profile seems fine, and nobody can find anything wrong with your womb or tubes.

'And as far as you're concerned,' he said, turning to the man, 'you seem to be producing perfectly good and active sperm in perfectly reasonable numbers.'

He paused, scanning further down their file. 'And there are no immune problems . . . you're using fertility predictor kits . . . and you're having sex every two or three days as well.'

'More than that,' said the man indignantly.

'Ah, yes . . .' He gave a short laugh. 'Yes . . . very good.'

He paused again – more for effect than because he didn't know what to say. 'Anyway,' he said with sudden purpose. 'I really don't see any problem. IVF can handle this. All we have to do is collect some eggs from you . . .' he looked at the woman, then at the man ' . . . and I'm sure you can give us some sperm without any help from us. Yes? Then all we've got to do is mix them together and put a fertilized egg back into your womb – I take it you want to carry the baby. You don't want a surrogate? No? We might need two or three goes, of course, but we're fairly efficient here. Any questions?'

If he speeded things along, he thought, he might get the chance of a break before seeing the next couple.

'Can we choose how many children we have?' asked the woman. 'We sort of thought – well, we wondered. Maybe we should have two if we can – because that's all we want, and it will save us having to go through it all again.'

'Yes, you can probably have two if you want. It will be cheaper for you in the long run. As long as we get more than one egg off you in the first place and as long as two get fertilized and look as though they're developing properly, then we can certainly put two back in, if you want. Years ago, we actually needed to put several back in, because so many didn't take, but we've solved this problem now, so you can have what number you want. We really don't recommend more than two, though, and most people are still better with just one. It cuts down on the risk of complications.'

The couple had no further questions, and the remainder of their time was spent discussing dates and times for the treatment to start – and methods of payment. He buzzed for Marilyn to show them out. Afterwards, she came back in to clear the table. Leaning over him to pick up his coffee cup, her long hair brushed against his cheek. He pushed back the chair to give her more room.

'Have you got any children, Marilyn?'

'Give me a chance,' she said, 'I'm only twenty-one.'

He shrugged. 'Lots of women do by your age. Don't you want any?'

'Sure, but not yet... Well, maybe.' She stood facing him, holding the tray. 'I suppose if I met somebody really attractive – or somebody really rich – who was prepared to support a child with me... Maybe I'd consider it.'

She didn't tell him that if *he* would consider it, she'd take him up on it like a shot. 'The trouble is,' she continued, 'I haven't met any man I'd like to live with. And those I fancy just for sex don't fancy me. And those that are rich, well, they think I'm just after their child support. It's not easy for a girl these days. Anyway, how about you? Do you have any children? Presumably you have.'

'Two,' he said, 'both with the same woman.'

'Do you live with her?'

He nodded. 'For the moment, anyway – though we're thinking of moving on.'

'Funny,' she said. 'I always thought you'd have lots of children.'

'I'm only forty,' he said. 'Anyway – I intend to, eventually. I just want to spread them around a bit. Have them with several women, you know?'

She knew very well. 'Shall I show in the next couple?' she asked, hovering near the closed door.

'Please.'

They smiled at each other.

Friday: 14.30

'I gather you're about to begin treatment for your erectile dysfunction,' he said to the timid, balding man in his mid-fifties. He loved saying those words. 'Why not wait and see if that works? Why rush to have a baby?'

'It's my fault,' said the man's partner. 'I'm desperate to have a baby and I'm not getting any younger. I'm thirty-five already. I've put it off and put it off – but now, just when I've decided I can't leave it any longer, this happens. I've told him I don't mind having it with somebody else – or even moving on. But he really wants to father my first. I think the whole thing is putting pressure on him. That's the problem. It's psychological.'

'It's not,' said her partner. 'I can't get a proper . . . even when I . . . when I. . . . Well, you know. I come while I'm still limp. There's something wrong with it. I know there is.'

'But you get perfectly good hard-ons in your sleep,' said the woman. Then, turning back to the doctor, she said, 'He even has wet dreams occasionally.'

The doctor decided it was time to interrupt. 'Yes, well, I'm sure your treatment will sort you out. But if you want to go ahead with IVF in the meantime, the only thing that really matters to us for the moment is how we collect your sperm. It says here that your impotence doesn't stop you from masturbating. I gather from what you've just said that you can still manage to ejaculate? Your dysfunction hasn't got any worse? So do you think you'll be able to produce some sperm for us when the time comes?'

The man shuffled uneasily. He just could not get used to discussing sexual matters so openly – especially when they made him feel so useless.

'He comes at half-cock,' said his partner, much more at ease with the matter, 'but he does still come.'

'Well, then, that's fine – half-cock is cheaper than no cock at all . . . as we say . . . in the trade . . . yes . . . well, what I mean is . . . it's cheaper for you if you *can* produce the sperm yourself. We can try artificial insemination first, if you like, or you can go straight to IVF. It's up to you. It's all a question of how much you want to pay, really. But you've always got the reassurance that, when the time comes, if you can't manage it, we can always get some sperm straight from your testes and inject them into the egg. It's more expensive and not as satisfying for you. But it does work, so you needn't feel too pressured about *having* to ejaculate when the time comes.'

After the couple had gone, Marilyn came in to collect their file.

She perched herself on the edge of his desk as he walked around the room, stretching his legs. The wrap-over part of her skirt gaped provocatively.

'Don't you think,' she said, 'that the more people you help to have children that couldn't manage it by themselves, the more trouble you're creating for the future? Isn't everybody going to end up infertile or impotent at this rate?'

'Does it matter?' he asked, trying hard not to stare at the exposed bit of her thigh. 'As long as the technology is there for people to reproduce, does it matter if they can't manage it by themselves?'

Before she could answer, they heard the next couple enter the reception room. Marilyn slid off the desk, adjusted her skirt, and went to greet them. On her way, she walked much nearer to him than she needed and paused while still within touching distance.

'Well, *I* think it matters,' she said. 'I like sex. And the thought of living in a society in which all men are limp just doesn't appeal to me.'

He laughed, but didn't respond. Instead, he asked for another coffee – and perhaps some biscuits – after he'd seen the next couple. 'Join me,' he said. 'Let's talk some more.'

Friday: 15.00

'So, you were both tube-blocked in your twenties...' he said, reading their file. 'A standard vasectomy for you...' he glanced at the man, a university lecturer in his early forties '... and a standard tube ligation for you,' he said to the woman, a well-known journalist, in her mid- to late thirties. He had seen both of them occasionally on television. 'And now you've decided it's time to have children.'

The couple nodded.

'Fine. Well, no problem. The only thing we've got to decide is whether you want us to open up your tubes again so you can try for yourself or whether you want to go straight to IVF. I see you didn't bank anything, though. Still, never mind. We can take some sperm out of your testes, some eggs out of your ovaries, and off we go.'

'Which would you recommend?' asked the woman.

'I would go straight to IVF,' he said. 'Then your contraceptive surgery can stay as it is. Much neater. And there's still no risk of unplanned pregnancies once you've had your family.'

Ten minutes or so later the couple left, arranging the dates to start their treatment on their way out. Shortly after, Marilyn came in with the coffee and biscuits. He invited her to sit in his comfortable chair while he walked over to the window and looked out over the city again.

'You've had a lot of tube-blocked couples in this week,' she said to him, kicking off her shoes and putting her feet on his desk.

He shrugged. 'Everybody's doing it,' he said. 'It's so simple. Stop sperm from ever being ejaculated and never let eggs reach the womb. There you are. No unwanted pregnancies – no abortions. Then use IVF to fertilize an egg and put it straight into the womb. What could be easier? Of course, when they were blocked it was early days – about 2035 – and blocking was still done by surgery. Very medieval. But it's still a great system.'

'I think you'd like to see all children conceived by IVF. I bet you'd do away with sex altogether.'

'I wouldn't do away with sex,' he said. 'That's not what we're talking about. I like sex as much as. . . . as . . . well, as *you* do. I'd just separate sex from reproduction, that's all. Everybody would have their plumbing blocked at puberty and reproduce by IVF.'

'But not everybody can afford it. IVF is so bloody expensive.'

'The price is falling,' he said. 'Besides which, it would pay the State to subsidize it. Just think. Every embryo would be screened for genetic diseases before implantation and every pregnancy would be planned. There'd be no more need for abortions, no long-term care of people with those terrible diseases – no infertility treatment, because fertile and infertile would be treated the same. It would save a fortune.'

Moving away from the window, he came and stood near her. Lying back in the tilting chair, feet still on his desk, her head was just about level with his waist. She put her hands behind her head and looked up at him, smiling mischievously.

'What about you? What did you do – for your two children? Have you been blocked?'

He recognized a leading question and thought quickly. 'No,' he said simply. 'We did it the old-fashioned way.'

Her expression became even more mischievous. 'You're not impotent either, then.'

Amused by her impudence, he shook his head in reply, grinning broadly. As he did so, the telephone rang in Marilyn's office. Patting him lightly on the chest as she stood up she said, 'I'm very glad to hear it,' then scampered in her stockinged feet to answer the call.

Friday: 15.45–17.00

Two more couples were interviewed. In the first, some childhood disease had blocked the woman's oviducts and the man was ejaculating live sperm but they were unable to swim. In the second, the woman was perfectly healthy, as far as all the tests could reveal, but there were no sperm at all in the man's ejaculates. Biopsy had shown, though, that his testes were beginning the process of making sperm, it was just that none ever developed fully.

'Easy,' he said to both couples. 'We'll take the most promising sperm cell we can find and simply inject it straight into an egg. No problem.' He told the first couple that all the man would have to do was produce a sample, as he had done for the tests. And the second couple that they would take a tiny part of his testes and get the cells – young sperm he called them – they needed that way. Expensive but effective, he told them.

Marilyn shook her head as she brought in the file on the last couple he was going to see. 'I bet some of these problems with sperm and testes and tubes are genetic,' she said.

'Some of them,' he replied. 'But a lot are due to infections – just bad luck.'

She sat on the edge of his desk again, while he stayed seated in his chair, relieved the afternoon's work was nearly over.

'So all you're doing,' she said, 'is breeding future generations that can only have children by IVF.'

He was confused. Her language was aggressive but her body language was seductive. She had positioned herself so that the slightest movement of his rotating chair meant that their legs touched.

'Not entirely,' he said. 'We know quite a lot about the genetics of infertility – which genes are responsible for which. The beauty of IVF is that, once we get the eggs and sperm in a test-tube, we can get rid of things – like genes for infertility – with a bit of genetic engineering. There's nothing we can't do, as long as fertilization takes place outside of the body.'

She didn't answer immediately, so he changed the subject.

'Anyway, it's your last day – though I hope we can get you back some time. Are you rushing off?'

'I don't have to.'

'While I'm dealing with the last couple, why don't you nip out and get a bottle of wine from across the road? Put it on my account. Get two, if you want. Let's have a private party – and you can tell me everything you don't like about IVF and how much you hate the job I do.'

Friday: 17.30

The last couple was very different from those who had gone before. The man was in a wheelchair. The lower half of his body had been crushed and his genitals had been so badly damaged they had been removed.

The consultant flicked through his file. 'We found a few bits of testicular tissue,' he said, 'but there was no sign of any sperm development. What we're going to have to do is to make some sperm for you. We'll take a cell from somewhere in your body – probably your chest – get rid of half the chromosomes from the nucleus, and inject it straight into the egg. It's as easy as that. Who needs testes?'

Friday: 18.10

'Did you get the wine?' he asked, taking off his jacket and tie as Marilyn tidied his office.

'Of course, two bottles like you said. They're in the fridge.'

He selected two glasses from the assortment hidden away in a small cupboard and walked into the reception room, Marilyn following close behind.

'Before we start,' she said, 'there's one place I'd like to see.'

'Where's that?'

She laughed, mildly embarrassed. 'I'd like to see the wanking room. I've often wondered what it's like, this place where all these men go to toss themselves off to order.'

He took the wine from the fridge. 'OK,' he said. 'I'll take these with us. It's probably the most comfortable room in the building, anyway. I think you're in for a surprise.'

They took the lift down two floors to the main laboratory area, then went along a carpeted corridor before stopping outside the only door that had no sign on it. He unlocked it and stepped aside so that she could go in first, then switched on the light.

'My God, there's a bed,' she said, walking in and bouncing up and down on its edge, testing the mattress.

'And an en-suite bathroom,' he said. 'It's better than most five-star hotels. There's even room service. We have an account with the all-night restaurant down the street. Security lets them in and out.'

'Why?' she asked, incredulously. 'No wonder it's expensive.'

'Well, for straightforward IVF, and especially GIFT, sperm are more fertile if they're collected during sex than from masturbation. So couples can book in here for an hour or so – or for a whole night if they prefer. It's surprising how many can't manage it though, once the pressure's on.'

Marilyn went over to the settee which was scattered with porno-graphic magazines. She flicked through a couple, then put them down.

'Are there any films?'

He pointed to the monitor. 'Take your pick.'

She held up her hand. 'Pour me some wine,' she ordered. 'Can I have a shower? Then we'll watch a film.'

She hesitated. 'Have you got time.'

He smiled. 'Actually, I've got all night.'

'Funny you should say that,' she said as she disappeared into the bathroom.

Saturday: 09.00

'There is one thing IVF can't do,' Marilyn said as the pair ate their breakfast. They were sitting up in bed, still naked. She finished her Buck's Fizz and poured them both a coffee.

'What's that?'

It was actually the first time they'd mentioned IVF since she'd had her shower. She *had* got dressed again before coming out of the bathroom – at least, she'd put on her shirt and skirt – but it had been only temporary. One – rushed – glass of wine and about five minutes of hard-core video was all they had needed to give them the excuse neither had actually required. They'd had sex twice, ordered a meal, had sex again, slept in each other's arms, woke for early-morning intercourse, then ordered breakfast. In between they'd talked at length about everything – except IVF.

'It can't give you a spontaneous conception,' she said. 'How many children do you think used to be conceived from relative strangers suddenly lusting after each other, even though they'd no past and no future?'

'I've no idea,' he said. 'But what I do know is that those were just the babies who were either raised by somebody who wasn't their father or who were brought up in poverty because the mother ended up on her own.'

'My God, how old are you? Ninety? It's hardly a problem, now, is it? Everybody knows who the father is and every baby gets its father's support. Don't you think that what we've just done is a much nicer way of starting a baby than arranging it six months in advance, seeing consultants, having tests, and so on?'

She hesitated, suddenly realizing what she had betrayed. 'How *would* you feel if I've just conceived?' she added tentatively. Not that it mattered how he felt, now – the deed was either accomplished or not. She knew better, though, than to tell him that her 'spontaneous conception' had itself been planned – from the minute she met him at the beginning of the week and her fertility predictor gave her the red light. The room, the bed and the room service – they hadn't really been a surprise to her. They were one of the first things she'd heard about from the other receptionists and secretaries when she'd begun to work there. She'd also known his income and how many children he was supporting. Even if he hadn't been comfortably rich with only two children, she might still have found him sexually attractive. It had all seemed far too good an opportunity to miss.

He nodded. 'I'd be happy,' he said. 'How would you feel?'

She pretended to think for a second, then nodded as well.

He began clearing the bed of the trays and remnants of breakfast.

'What are you doing?' she asked. 'I haven't finished yet.'

'Don't you think we should do it spontaneously just one more time?' he said, pulling back the bedclothes in preparation for kissing her body. 'Just in case it didn't work the first few times.'

He began to kiss and stroke her. This was not the moment to tell her he'd lied about how he'd had his first two children. He *had* been blocked – about ten years ago. His children *had* been produced by IVF – and so would any future children he had. Several times over the years he'd had reason to be glad he couldn't produce children simply by having sex; so often he had been the target of women seeking to lay their hands on his child support. As it was, he'd been able to enjoy himself with them all – and still stay solvent.

As he slipped into position to enjoy himself one more time, he gave a silent toast. Here's to IVF, he thought.

Every Man Fertile

At about the year 2050, this scene is set a little further in the future than Scene 3, the one involving surrogacy. In the meantime, the success rate of IVF has risen so that a baby is almost guaranteed after two or three treatment cycles. The problem of multiple births has also been solved. At least some wealthy men have discovered how to avoid being seduced into child support. And, most importantly of all, there is no such thing as male infertility. Every man – even one without testes – can produce his own genetic children.

To the modern man – as to the modern woman – infertility is one of the most puzzling and often distressing legacies of our biological past. Help is already to hand, of course. The vast majority of infertile men can now find hope in either artificial insemination or *in vitro* fertilization (IVF) where none existed before. And technologies only just around the corner will mark the end of infertility for men just as surely as they will for women. Yet, as this scene has just hinted, the impact of assisted conception on human reproduction has scarcely begun. Although reproductive technologies all originate as cures, they are destined, without doubt, to metamorphose into commodities.

Before considering the future, though, let's first consider the modern treatment of male infertility.

Impotence and Artificial Insemination

A man's reproductive system, like that of all male mammals, consists of three main parts: a pair of testes to produce sperm; various glands – such as the prostate and seminal vesicles – to produce seminal fluid; and a penis with which to inseminate the entire mixture into the female's vagina. Running between and through these different parts is a plumbing system of tubes that join the whole lot together.

In the past, most couples have relied on the man being able to achieve an erection in order to get the whole conception process started. If he could not, then the couple could not have children. Normal erections are produced by a sudden influx of blood to the penis, which hence becomes stiff. The influx occurs when certain muscles relax, allowing blood to enter via the arteries, while veins are pinched tight, blocking the blood's exit. Problems with either exit or entry of the blood can cause either impotence (blood entry never being faster than exit) or priapism, a painful, extended erection (blood entry being greater than exit for long periods). Both can lead to difficulties with intercourse, although of the two impotence is the clearer problem.

Impotence

Impotence, or penile erectile dysfunction, is a common medical condition that can affect up to half of all men in their fifties and sixties. Roughly 10 per cent of men – which in the United States translates into a total of over 10 million men – are suffering from impotence at any one time. For many, particularly younger, men the condition may be temporary, associated with periods of stress. For some, though, the condition can be permanent. It is often distressing and is frequently associated with depression and poor self-esteem.

Throughout the 1970s it was thought that psychological problems caused 90 per cent of impotence. In the 1990s, though, it was discovered that most cases of impotence in men stem from physiological rather than psychological disorders – and now only 10 per cent of impotence is thought to be psychological. The favoured theory is that impotent men are unable to produce the chemical – nitric oxide – that triggers the penis to engorge with blood.

Nitric oxide relaxes a spongy muscle called the corpus cavernosum that runs down the core of the penis from the tip to the back. When the penis is flaccid, the muscle is in its contracted state and excludes blood from sinusoids, which are chambers in the blood system of the penis. On sexual arousal, nerve signals from the brain trigger the production of nitric oxide in the corpus cavernosum. Blood is then admitted into the sinusoids, which fill up and engorge the penis. Then, if a man with an erect penis ejaculates – or his interest in sex simply wanes – the brain stops sending the messages that produce nitric oxide. Once this happens, the corpus cavernosum contracts, the blood drains away and the penis returns to its flaccid state.

For years, researchers have been trying to develop a user-friendly and efficient therapy for impotence – preferably either a pill that can be taken orally (Viagra is the most recent) or a cream that can be applied directly to the penis. In the meantime, impotent men were normally helped to achieve erections via less friendly devices such as inflatable prostheses and injections of drugs such as prostaglandin E1.

When it comes to reproduction, though, as opposed to sex, there is already an alternative for impotent men: artificial insemination.

Artificial Insemination

Artificial insemination is one of the oldest forms of assisted conception. Essentially, it is the process by which sperm are collected either from a woman's partner (AIH – artificial insemination by husband) or an anonymous donor (AID – artificial insemination by donor) and introduced by syringe or other appropriate device into the woman's genital tract. The process has been with us for two centuries, ever since the Scottish barber-surgeon John Hunter successfully inseminated the wife of a linen draper, using her husband's sperm, in 1790.

Although artificial insemination has been part of the human scene for at least two centuries, until recently the process was mostly used for other animals – particularly cattle, horses and recently pigs. Sperm are collected from one – or several – males. Then, more often than not, they are frozen, to be thawed later and used to impregnate females. Nowadays, human sperm are also often frozen and stored for later insemination. Use of frozen sperm has about two-thirds the success rate of freshly collected sperm at producing pregnancy – about 10 per cent of women conceive per menstrual cycle using frozen sperm.

Clearly, artificial insemination is most helpful to a couple when a man can ejaculate during masturbation but cannot penetrate – as in the second couple in the scene. It is also appropriate in cases where a man has 'banked' sperm in advance of major operations with a risk of subsequent infertility. It is less satisfactory if the man cannot ejaculate at all, or if his ejaculates are infertile. Then, if the couple opt for artificial insemination, their only choice is AID. But whereas this helps a woman to produce her own genetic child it does nothing to help her partner.

Not surprisingly, artificial insemination has raised a number of concerns, though most have now subsided. An initial worry was that it might lead to an increase in birth or developmental defects. Perhaps bypassing intercourse, particularly with sperm that have been frozen for some time, could have a deleterious effect on the child conceived. Maybe a woman's body would be less rigorous than usual in selecting 'fit' sperm from the initial ejaculate. Or maybe the handling of sperm and their exposure to chemicals on glassware, syringes and so on, could cause damage to their DNA. And as for freezing sperm, maybe the low temperature could also damage the sperm DNA. In fact, there is no evidence in humans – or livestock – that the use of either fresh or frozen sperm creates an increased risk of birth defects. Nor is there any evidence that children conceived by artificial insemination are at any greater risk of psychological or social problems.

Artificial insemination raised another question that became an amusing diversion for most people but was deadly serious for some. Ethicists of various persuasions debated whether virgins should be allowed to conceive. The debate was at its height around 1990. In truth, any outlawing of virgin births would have been fairly unenforceable since women can artificially inseminate themselves with just a cake-icing syringe and an obliging – but distant – male acquaintance. Nevertheless, a variety of psychologists, psychiatrists, politicians and editorial writers proclaimed that they weren't happy with the idea, and various luminaries from the Christian Church, such as the Archbishop of York in Britain, seemed to fear the loss of a historic monopoly. Nearly ten years on, society seems to have survived the virgin onslaught.

Of greater concern to most people was the possibility of eugenics and inadvertent incest that artificial insemination raised. As long as the process was limited to established couples in which the man had erectile problems, such fears were minimal. But in the 1970s, with an increase in AID, and in the 1990s, with an increase in demand from single women, both problems gained a higher profile. Increasingly, women sought sperm from famous men – Nobel Laureates, for example – and increasingly, licensing authorities lost track of who was related to whom.

We shall discuss the questions of eugenics and incest in later chapters, for both will become central issues in the reproduction of future generations. For the moment, we concentrate on the treatment of male infertility. In fact, artificial insemination is unlikely to be a major process in future reproduction. Its place will be taken by IVF, potentially a much more universally useful method.

IVF – a Helping Hand for Sperm

Of the one in six couples who find it impossible, or at least extremely difficult, to conceive, between 30 and 50 per cent have problems that can be traced to the male. IVF was initially developed to cure female infertility. Increasingly, though, it is being used to cure male infertility. And if men are not infertile because of impotency, they are infertile because something is wrong with their manufacture or transport of sperm – and IVF can help.

As with women, many things can cause male infertility. Hormonal or structural abnormalities, excessive use of alcohol and drugs, and illness – particularly urinogenital disease – can all be to blame. Problems can arise at any point in the reproductive process. Just how IVF can be used to help depends on the symptoms of infertility, and the most recent technologies have been developed to cope with ever more severe symptoms. This section is organized around the development of IVF technology, not around the development of sperm. As mirrored by the sequence of couples in Scene 4, therefore, the following discussion in effect works backward through sperm development. We begin by considering problems with the ejaculate and end with infertility generated by problems at the very earliest of stages in sperm development.

Men with a Low Sperm Count – Standard IVF

Some men habitually have fewer sperm than expected in their ejaculates and as a result are often infertile in normal intercourse.

But because standard IVF procedures require relatively few sperm, the process can be used as a way of helping such men, and increasingly, this is what is happening.

In standard IVF protocol, a man produces an ejaculate by masturbating, then sperm and eggs are mixed together in a Petri dish for fertilization. As IVF was originally developed to treat female infertility, one of the earliest requirements for a couple to be accepted for IVF treatment was that the man should be producing as fertile an ejaculate as possible. His suitability was judged by asking him to abstain from ejaculating for three days or so, then to produce an ejaculate by masturbating. Collected under such conditions, an average ejaculate would contain about 300 million sperm. At least 60 per cent of those sperm would look 'normal' and well over half would swim actively. More often than not, these characteristics would indicate a fertile ejaculate. Nevertheless, many an IVF procedure failed because for some unknown reason none of the eggs were fertilized by the man's sperm.

One potential problem, as mentioned in the scene, was that the sperm were collected via masturbation. Sperm fertilize eggs more proficiently during IVF and GIFT procedures if they are collected during intercourse in special condoms rather than during masturbation. This is not surprising. Natural selection shaped masturbation as the means by which a man rids himself of old and unwanted sperm – hence the urge to masturbate after a period of abstinence. At present, though, the sheer convenience of masturbation means that this remains the usual procedure for collecting sperm. Maybe in the future, with clinics competing for customers, things will change and executive-style collecting rooms as described in the scene will become commonplace.

Even if an ejaculate is fertile, the concentration of sperm around the eggs in the Petri dish is critical to the success of IVF. There is an optimum – not too many sperm, and not too few. Surprisingly, less than 1 per cent of sperm are fertile *at any one time*. Enough have to be added to the Petri dish, therefore, for there to be a reasonable chance of at least one fertile sperm encountering each egg. If too few sperm are added, some eggs may not be fertilized. The solution, however, is *not* to add all the sperm a man can produce.

If too many sperm crowd around an egg there is a danger that more than one sperm will fertilize the egg simultaneously. Normally, having embraced one sperm, an egg passes a chemical message across its surface and, within seconds, becomes impenetrable to any later sperm that might arrive. If two or more sperm gain access before the egg has time to set up its protective barriers – a phenomenon known as polyspermy – the egg will die just as surely as if it were not fertilized at all.

IVF clinics differ in how many sperm they add to their Petri dish per egg, but it is usually of the order of about 25,000. So, if ten eggs were harvested from the woman, 250,000 sperm would be added from the man's ejaculate. This might sound like a lot of sperm but it is only about one for every 1000 sperm the average man will have ejaculated. Normally, the excess would be discarded.

It is because IVF requires relatively few sperm that standard IVF procedure can be used to help men whose ejaculates contain few sperm. Increasingly, therefore, through the 1980s couples were accepted for IVF treatment in which the problem lay with the man rather than, or as well as, the woman.

The main reason for men to ejaculate too few sperm is a blockage in the tubes – the epididymides, vasa deferentia and urethra – that carry sperm from testes to penis. After being born and grown deep in the testes, sperm enter what is in effect a maturation and storage tube – the epididymis – which clings to the outer surface of the testis, just under the scrotal skin. There are two epididymides, one on each testis, and they are positioned so that the stored sperm – which are damaged by undue heat – can stay as cool as possible. Each epididymis runs from the top to the bottom of the testis and is incredibly convoluted.

Once the sperm are in the epididymis they are, in effect, simply queuing and waiting to be ejaculated. They are on a conveyor belt. Each time the man ejaculates sperm from the front of the queue, the rest shunt forward. Newly mature sperm are joining the back of the queue all the time. Very approximately, each sperm will spend about two weeks queuing in an epididymis.

While queuing, the sperm complete their maturation before beginning the final part of their journey to the man's penis. At the bottom of the testis, the epididymis rather suddenly straightens out and becomes the vas deferens. It is through this part of the tube that the sperm will finally be shunted. Again, there are two vasa deferentia, each one eventually joining the man's urination tube – the urethra – which runs from his bladder, through the shaft of his penis, to open to the outside at the penis tip. From the epididymis, each vas deferens first runs up the back of the testis. It then enters the man's body cavity near the base of his penis, loops up over a tube running from the kidney to the bladder, and finally runs down by the side of the bladder to join the urethra. At the place where the two vasa join the urethra, just where it leaves the bladder, they pass through a central walnut-sized gland, the prostate, which produces the bulk of the seminal fluid. Sperm may spend up to five days or so in a vas deferens, depending on how long the man goes between ejaculations. They don't enter the urethra, though, until a few moments before ejaculation.

Most of the time a man has no sperm in his urethra. Even during the early stages of masturbation or intercourse, when his penis becomes erect, his urethra is largely empty of sperm. Beyond a certain level of sexual excitement, though, sperm are shunted out of each vas deferens and into the urethra. A round sphincter that normally prevents urine leaking out of the bladder now prevents sperm from entering the bladder. The man's urethra is loaded, ready to fire. Eventually, seminal fluid pours from his prostate into his urethra. Then muscles contract and the mixture of fluid and sperm is projected in a series of spurts along the urethra and out into the woman.

Several types of system failure can reduce the number of sperm in a man's ejaculate. A weak bladder sphincter, for example, can result in many sperm entering the bladder rather than spurting from the penis. Most problems, though, are caused by a blockage somewhere in an epididymis or vas deferens. The number of sperm ejaculated can be halved if the whole supply of sperm from the testis on one side is cut off. A blockage on both sides can reduce the number of sperm in the ejaculate to zero. This is, after all, the principle of vasectomy in which both tubes are cut or ligated (tied) as a form of contraception. Natural blockage of both sides is particularly likely if the prostate gland becomes infected, because both vasa pass through it and can be occluded, either by pressure or scar tissue.

Sometimes natural blockages simply cause sperm to queue for extended periods, so that by the time they get past the hiatus, they are old and relatively inactive. The job of removing old and dying sperm from vasa blocked by disease or vasectomy falls to white blood cells. These are usually present in the epididymis and vas in only small numbers, but can multiply enormously if there are many sperm to be removed. Large numbers of dead and dying sperm and their accompanying white blood cells can sometimes trigger immune reactions that might themselves harm a man's health and fertility.

Blockages in the sperm transport system can occasionally be corrected surgically, bypassing the blocked stretch – just as vasectomies are often reversible. In many ways, though, IVF is a simpler treatment for men with low sperm numbers because so few sperm are needed.

If a man ejaculates no sperm, however, standard IVF cannot

help. Such men had little hope of fathering a child – until the 1990s and the advent of TESE.

Men with Blocked Tubes – Testicular Sperm Extraction (TESE)

Some men have plenty of sperm in their epididymides but no sperm in their ejaculates simply because their vasa deferentia are totally blocked. The answer is astoundingly straightforward: take sperm directly from the testes and fertilize the egg using IVF.

Surprisingly, sperm actually become *less* fertile when they are ejaculated. Some first become capable of fertilizing an egg via IVF almost as soon as they enter the epididymis. As the cohort moves along the queue in the epididymis, more and more of the sperm become fertile. The most fertile part of the queue is that near the bottom end of the epididymis, just as they are about to enter the vas deferens. Then about one in 200 of the sperm would know what to do if they bumped into an egg in a Petri dish. But as soon as the sperm are engulfed in seminal fluid and inseminated into a woman, even these one in 200 lose their fertility – temporarily.

It is thought that the seminal fluid coats each sperm with large molecules – a sort of protective cloak – in preparation for its journey through the female. As the gangs of fertile sperm make their way through the woman, the large molecules on their surface are gradually stripped away in a process called capacitation, and many of the sperm once more become capable of fertilization.

In standard IVF, sperm are encouraged to capacitate by chemical treatment during washing and incubation. Whatever the treatment, though, sperm collected from the epididymis are as fertile as – if not more than – the sperm collected from an ejaculate. It was partly for this reason that researchers in Missouri and Brussels developed the procedure known as Testicular Sperm

Extraction (TESE). Even if a man has totally blocked sperm tubes, enough young, healthy and fertile sperm can often be collected from the testis side of the blockage and used for IVF. Of course, TESE is invasive, expensive and a much less satisfying procedure for a man than providing sperm via ejaculation. Nevertheless, for men with simple blockages, it is a much-needed solution.

TESE gave hope to one of the commonest categories of infertile men – those for whom past or current infections had simply blocked their sperm transport system. It also gave hope to vasectomized men who had changed their mind and now wanted to father a child, or a further child, but who found their vasectomy could not be reversed. It did not, however, give hope to men who had very few and/or relatively sluggish sperm in their ejaculate. Even the helping hand of standard IVF could not assist these men who had to await the development of ICSI.

Men with Few and/or Impotent Sperm – Intra-Cytoplasmic Sperm Insemination (ICSI)

Increasingly in the 1990s modern variants on IVF procedures concentrated on the treatment of those cases of male infertility in which the man produced few or no sperm – or in which the sperm that he did produce were lethargic or chemically impotent. All these problems were linked in that they could be cured by the micromanipulation of sperm and eggs. But what could cause men to be infertile in such a desperate way?

Fertile men's testes are a hive of activity. Inside, cells are multiplying, growing and finally maturing into sperm. Normally, the rate of manufacture is incredible – around 300 million sperm a day during the years of peak production between the ages of twenty and thirty. That means that well over 1000 sperm are matured with every beat of the man's heart. This does not mean, though, that each sperm is made quickly. In fact it takes about seventy-two days – over two months – from first cell division to a sperm being ejaculated.

The cells that produce sperm are known as stem cells. There are millions of these inside a testis and their job is to divide to produce cells that, after further multiplication and development, mature into an adult sperm. Stem cells are virtually immortal, continuing to divide throughout a man's life, and are interspersed and associated with specialized nurse cells that nurture and support them. Throughout the testes, stem cells line the walls of tiny tubules – the so-called seminiferous tubules. As young sperm are produced, they are shed into the middle of these tubules and gradually move along them. They are shunted step by step into increasingly larger, longer and fewer tubes until they reach the epididymis.

Some men have few sperm in their ejaculate because they *manufacture* very few sperm, not because they have blocked tubes. Maybe their stem cells fail to divide, or divide only infrequently. Alternatively, the process of division, growth and maturation may begin normally, but then arrests at some stage before mature and fertile sperm have been produced. Sometimes, only some of the seminiferous tubules are manufacturing sperm capable of maturing. The result is that neither the man's ejaculate, nor material taken from his epididymis, contains enough fertile sperm for success in a Petri dish.

This is one scenario in which a man's sperm need more help than simply being placed in the same dish as an egg. There is another – and it isn't necessarily associated with low sperm numbers. It is when sperm for some reason – chemical, structural or behavioural – are incapable of penetrating the egg even when they bump into it.

An egg arrives at the fertilization zone like a fortress, surrounded by three lines of defence that have to be breached before the egg will yield. The outer line of defence, the cumulus, is a thick layer of cells that the egg has brought with it from the ovary. Beneath the cumulus is another relatively thick, smooth layer, the zona pellucida, which is the outer membrane of the egg itself. Under the zona is a narrow space that surrounds the final, most vulnerable barrier, the vitelline membrane.

Using their heads, fertile sperm hack their way through the cumulus cells. If successful, and they reach the zona beneath, they stick the side of their head on to the membrane with chemicals. With this chemical attachment as an initial purchase, the sperm again uses its head to cut a way through, this time using a pointed spike that has been exposed at the tip. The lashing tail provides the force to push the sperm forwards. Finally, if a sperm is the first to get through the zona, cross the underlying space and touch the vitelline membrane, the egg engulfs it in a welcoming embrace. The successful sperm sheds its membranes within the egg, releasing its genetic heart of DNA. This then travels to fuse with the similar heart of the egg.

Even if a man's testes are producing sperm, they sometimes never develop the ability to fertilize eggs. Maybe they cannot generate enough power with their tail to breach an egg's defences. Or maybe the chemistry of their head and membranes bars them from penetrating the various layers. Some men's sperm do not capacitate, in which case a sperm does not shed the cap of chemicals on its head to expose the pointed spike. Shedding the cap is an essential precursor to fertilization and usually happens when the sperm is making its way through the outer layers of the egg. If a sperm hasn't shed its cap by the time it reaches the zona pellucida, it is incapable of travelling that last short step and completing the final act of fertilization.

Whether a man is producing few sperm or impotent sperm, his infertility can now be treated by making sure that even if there is only one sperm, and that sperm is impotent, it still penetrates the egg. The techniques developed include drilling, tearing or chemically burning holes in the egg's outer coatings –

or using a needle to inject sperm directly into the egg itself. The earliest such technique was called SubZonal Insemination (SUZI) – the injection of sperm into the perivitelline space under the egg's outer coating. The most recent, though, known as Intra-Cytoplasmic Sperm Injection (ICSI), involves injecting a single sperm right into the cytoplasmic heart of a mature egg.

ICSI was introduced in 1992, thanks to the amazing success of a team at the Free University of Brussels. Unhampered by government regulations, the Belgian team was able to launch into treatment with minimal data on safety. Their claim that 50 per cent of the eggs were successfully fertilized and that these eggs were just as likely to establish a pregnancy as those fertilized by conventional IVF had clinics around the industrialized world clamouring for licences. ICSI makes light of the problems created by small numbers of sperm in the ejaculate or epididymis. In principle, only one sperm is needed from the man – and that sperm can be totally impotent. There was some initial concern over which sperm should be chosen when there was more than one available. But when it was discovered that fertilization and development could be achieved as long as the sperm had a nucleus, the technological problems eased even if other concerns increased.

The combination of ICSI and TESE solves the problem of infertility for 98 percent of men. In the remaining 2 per cent of infertile men, which represents about 8000 men in the United States, there are no mature sperm, even in the testes. For these men, although sperm development in the seminiferous tubules begins, it arrests and no mature sperm are ever produced. Even these men, though, can now be helped. The very latest addition to the suite of IVF technologies doesn't even need a mature sperm.

Men who Produce No Mature Sperm – Round Spermatid Injection (ROSI and ROSNI)

In 1993 Japanese researchers discovered – first for hamsters, then for mice – that round, tail-less male sex cells that have not yet developed into sperm could be used to fertilize eggs by injection and produce normal offspring. The implication was clear – ICSI had the potential to help even men who manufactured *no* mature sperm.

Scientists had long wondered at what point in sperm formation the male sex cells became capable of fertilizing an egg, although we have to be careful here over what we mean by 'capable'. In normal usage, a male sex cell is only capable of fertilization if it can swim through a woman's cervix, surfboard through her womb, swim along her oviduct, and then inveigle itself into the very heart of her egg. And to do all of that, it needs to be a mature and potent sperm. In the current context, though, a male sex cell is considered 'capable' if, on being injected into an egg, it shares its DNA with the egg's DNA well enough for the union to grow into a healthy person. So, at what stage of development *does* the male sex cell become capable? To answer this question, we need to look at the cell's early life in a little more detail.

As we have just seen, a sperm begins its life as a cell – a stem cell – in the wall of a tiny tube called the seminiferous tubule. Each stem cell spends its entire life dividing and budding off other cells. Each of these cells divides several times, multiplying in number. Then they all grow, each developing into what is known as a primary spermatocyte.

Up until this point, every developing sex cell is just like every other cell in the body in that it has two sets of chromosomes, one from the man's mother and one from his father. When a primary spermatocyte divides to form a secondary spermatocyte, though, it undergoes a process – known as meiosis – that halves the number of chromosomes it carries. Meiosis not only halves the number of chromosomes each cell contains, it mixes up the DNA from mother and father so that each secondary spermatocyte contains a random mixture of the genes from the man's parents.

Each secondary spermatocyte divides again to produce the penultimate stage of sperm development, the spermatid. Spermatids still look nothing like sperm. Like all the cells that gave rise to them, they are round. It is when the spermatids begin to mature into sperm that they are shed into the hollow centre of the seminiferous tubule and begin their journey through the plumbing system of the man's testes.

The process of meiosis, by which the developing sex cells halve their complement of chromosomes, is of course an important step in the reproductive process. It means that when the mature sperm fuses with an egg, carrying a second half-set of chromosomes, the fertilized egg will have a complete set. The round spermatids are the first stage after meiotic division and so the earliest that can be injected into an egg without producing a zygote with too many chromosomes, which would be unable to develop. But in the Japanese experiments on mice, round spermatids *were capable* of fertilization – and shortly afterwards labs in Japan and other countries reported births in hamsters and rabbits. Scarcely months later, the first human pregnancies in 1995 and then births in 1996 were reported. An end to male infertility was in sight.

There are actually two techniques being used – ROSNI and ROSI. ROSNI is an abbreviation for Round Spermatid Nuclear Injection. It refers to the process of extracting round spermatids from men who are incapable of manufacturing mature sperm. These round spermatids are then lysed – dissolved from the outside in – and their nuclei used to inject into eggs. ROSI is a similar procedure but injects an intact round spermatid into the egg, not just its nucleus.

A few technical problems still remain. For example, many of the men who fail to develop sperm do not even produce spermatids. Even for these men, though, there is already hope. Work is underway to try to induce meiosis in spermatocytes in culture, and then use the resulting spermatids to fertilize an egg. If this succeeds, then as long as a man has primary spermatocytes in his seminiferous tubules, genetic fatherhood is still possible.

The final step on the road to universal male fertility will be to help men like the wheelchair occupant in the scene who did not even have spermatocytes – because he had no testes. Without doubt, though, it will not be long before any cell can be induced – or forced – to undergo meiosis, or at least to halve its chromosome content. Then such a *haploid* nucleus could be obtained from any cell in the body and used to fertilize an egg – and male infertility would truly be a thing of the past.

The Worries

Naturally, ICSI and its offshoots have raised all the concerns that conventional IVF has done, plus many more of their own. It will be a long time, though, before it is known whether children born following ICSI have any special problems. Fears are high that ICSI risks passing on genetic abnormalities to future generations because it involves forcing a single sperm directly into the egg.

There are also fears that the injection process itself might cause some damage. The glass 'needle' could harm the sperm or egg, or introduce 'foreign' substances such as bits of the egg's outer coating or even infectious agents into the egg. The primary concern, though, is over sperm quality and selection. In ICSI, a person selects the sperm to be injected; in conventional reproduction, the woman's body does a major part of the selection.

And thanks to natural selection the woman's body presumably makes a better job of judging what is a 'good' sperm than might a laboratory technician armed only with a microscope.

To see why, let's briefly consider the traditional selection process. This bars over 99 per cent of the sperm a man ejaculates from ever reaching a woman's oviducts. First, a woman gives the sperm in her vagina only a short period of time in which to save themselves – the length of time that her cervix is dipped in the seminal pool. On average, half of the sperm the man ejaculated fail this first step. Second, the sperm that do manage to escape the seminal pool and enter the cervical mucus are reprieved but not safe. Once in the cervical mucus they are quickly attacked in what looks like a second phase of quality control. Their lethal assailants are white blood cells unleashed from the walls of the woman's womb within minutes of insemination. As they advance through the cervical mucus, these relatively huge killer cells engulf and digest live and dead sperm alike. At their peak, the white blood cells can match the numbers of sperm but within twenty-four hours of insemination the hordes have gone, leaving behind much smaller numbers to complete the mopping-up operation.

The woman's army of white blood cells might look like a formidable and indiscriminate killing machine, but it is probably not. There are indications that the army might be very selective over which sperm it kills and which it spares. Some sperm seem to be immune to attack. Precisely what is going on is unclear, but one suggestion is that the white blood cells can tell the difference between sperm that carry genetic abnormalities and those that do not. Sperm free from detectable abnormalities are allowed through – and the remainder are killed. Even those sperm that do survive to meet an egg still have the final hurdle of fertilization when further selection for sperm quality can take place.

ICSI removes all elements of competition and selection from the sperm. Any defective genes carried by the injected sperm will be passed on to the child. Other *in vitro* methods, even standard IVF, allow at least *some* 'natural' contact between sperm and egg, making it more difficult for unhealthy sperm to achieve fertilization. But ICSI removes even this final barrier.

There is an additional worry associated with ICSI. Most cases of male infertility do not have a genetic basis – as we should expect after millennia of natural selection. As Marilyn said in the scene, though, modern developments in the treatment of infertility are poised to change all of that. Men who need ICSI to overcome their infertility – and particularly those who need ROSI – are much more likely to be infertile for *genetic* reasons than men who need only standard IVF. It has recently been discovered, for example, that a gene is missing from the Y chromosome in 20 per cent of men with very low sperm counts. Deletions in this gene, called the azoospermia factor (AZF), could be transmitted by ICSI and would probably result in similarly infertile sons.

The first few hundred babies born using ICSI in Brussels where the technique was pioneered showed no higher an incidence of birth defects than babies conceived naturally. But fears remain that genetic disorders might only become apparent long after birth, in childhood or adulthood. The most recent study suggests that ICSI babies *might* be at greater risk of heart defects and cleft palates, but still the actual risk seems relatively low.

Unlike most medical research, ICSI was introduced into clinical use without thorough testing in animals. Only long-term monitoring will show whether sperm injection really does or does not put the children it creates at a significantly increased risk. In the meantime, infertile men everywhere cross their fingers and hope.

IVF – the Contraceptive?

We can now see that, technologically, Scene 4 contained relatively few futuristic elements. Most of the couples that passed through the main character's clinic were being treated for infertility. Only the man in the wheelchair required an as yet non-existent technology to be able to gratify his ancient urge to reproduce. Running throughout the scene, though, was a great irony, and a signal of the way human reproductive life is likely to change over the next few decades. This theme will be developed further in later chapters, but the paradox can be noted here.

IVF and its associated technologies began life as cures for female infertility, expanded to become cures for male infertility, and are now poised to become routine commodities for everybody that can afford them. The irony is that although in their infancy the role of these techniques was to help people *make* babies, in their more mature years their most frequent role is likely to be as part of contraception – to help people *avoid* having babies.

The exciting – or should that be nightmarish? – possibilities are explored in later chapters. The important point here is that infertility – both male and female – will scarcely exist by the dawn of the twenty-first century.

5

Surrogate Testes and Ovaries

SCENE 5

One Man and his Rats

'Stop splashing like that – you're soaking me,' said the man, as his two children did their seal impressions. It was Thursday – his night to carry out the daily routine of bathing and story-telling while his partner went out for a couple of hours and took a break from motherhood. It was his two-year-old daughter doing most of the splashing, laughing loudly and fixing his eyes with hers, daring him to try and stop her. But his son, four years older, was taking full advantage of her misdemeanours to enjoy himself as well. Unable to stop the mayhem without losing his temper, the man opted instead simply to pull the plug and tell them it was time to get out.

'You first,' he told his son.

The young lad clambered out and stood dripping and laughing as the man towelled him dry. 'Now, stand still – I want to examine you.'

Used to the weekly routine, the boy stood still as his father placed his fingers under his scrotum and gently began to touch his testes. 'Open your legs a bit more,' said the man as he gently touched, prodded and squeezed his son's genitals, first underneath, then forward to the front, under the by now stiff little penis. He paused a second longer on his son's left testis, rolling it gently in his fingers like a marble.

The man knew he was being silly. The chances of finding a lump in a six-year-old boy were remote. But over the last few decades, the incidence of testicular cancer had soared and younger and younger males were being afflicted. He had been only twenty when, trying to find the source of the pain he had been experiencing for nearly a year, he found the deep-rooted lump in his own left testis. If his son was going to be afflicted, he wanted to make sure it was discovered as soon as possible.

'Did that hurt at all?' he asked. His son shook his head. 'OK, go and get ready for bed while I deal with your sister.'

Half an hour later, with his precious son tucked up in bed and listening to nursery rhymes, the man carried his daughter down the stairs. He knew there was no point trying to put her to bed yet. It would be at least two more hours before, in the midst of running around and creating havoc, she would suddenly lie down and fall asleep – maybe on the settee, or maybe just on the floor. Then he could carry her to bed.

'Shall we go down and look at the rats?' he asked, in the singsong voice he reserved specially for her. She had just learned to say 'Yes' and said it so beautifully he couldn't resist asking her questions just to hear her say the word.

Once in the cellar to his large house, the man opened the door to his rat-room. Even though the extractor was working full-time, there was no mistaking the smell, nor the scuttle as a dozen rats sought shelter while they waited to identify the intruders.

The whole room was given over to the rats. Not single isolated cages, but a grand and intricate maze of transparent Perspex tubes and caverns. Some tubes ran horizontally, some vertically. Nor were they confined to the walls. Scaffolding carried the tubes across the room at several different heights. As the man never tired of telling visitors, there were fifty potential nesting chambers and over 300 metres of piping. Even so, careful design allowed him to walk around the room unimpeded.

He liked rats, as well he should, and so did his children, as well *they* should. They would spend hours in the rat-room watching the bewhiskered traffic as the rodents went about their business, running from chamber to chamber through the transparent tubing.

'Shall we go and see the mummy one?' he asked his daughter, perched on his hip.

'Yesssss,' she sang, beautifully, nodding her head vigorously, then pointing in the direction they should go. She knew what he meant.

They paused for a while outside a section he had isolated from the rest of the colony. Inside they could just see the pregnant rat curled up, apparently asleep. His young daughter pointed excitedly. The man smiled. He wondered if he would ever tell her – or if he would ever tell anybody.

Two hours later, his daughter was asleep on the settee, doll by her side, the pair covered with a thin blanket. He stared at her face, beautiful in sleep, and marvelled that he had been given the chance to have such fantastic children. Rats were wonderful.

The months after he first discovered the lump in his testes had been purgatory. First there had been the weeks agonizing over whether to consult a doctor. He had read avidly, consulting the Internet and libraries, seeking reassurance. Testicular cancer was on the increase, but still the chances of his having it at twenty were small. For a while, he convinced himself the lump was just a swelling and would go away. But as it grew, and the pain intensified, he knew he was only fooling himself, and the longer he left it, the worse his chances would be.

When at last he went to his doctor, his fears were quickly confirmed – it was a growth, and quite a large one, and it could be cancer. A biopsy was needed, but if it was cancer they would need to act there and then, while he was still unconscious. He had decisions to make and needed counselling, both before and after.

He fell in love with his counsellor the second he saw her, even before they discussed what size testes he would like after his operation – if the worst came to the worst. They also discussed whether he would want children in the future, whether he would particularly want to have them via intercourse – and, if so, how he felt about the rat option.

He already knew about the rat option – who didn't? The initial public outrage nearly a decade ago had subsided but the treatment was still hotly debated. Every so often the 'it's not natural' lobby received an airing in the media, one or other politician would seek his or her moment of fame by vigorously supporting one side or the other, and then the debate would subside for another year. In the meantime, according to his counsellor, more and more men were taking the option and the first children born as a result were now alive and well and growing nicely. Try as she might to be neutral, it was obvious to him that she was a supporter of the option.

He found reassurance in the fact that as attractive and intelligent a woman as she could consider a man who took the option both as a lover and as a father of her children. He knew this – because he had asked her. And it was her answer that convinced him of what to do. Yes, he would bank some sperm as insurance. Yes, he would freeze some stem cells, just in case. But yes, he would also try to keep alive his chances of having children by intercourse in the future. No, he didn't want to draw too much attention to his balls by making them too big. And, yes, he would take the rat option – if the worst came to the worst.

The worst did come to the worst. It was cancer. He went under the anaesthetic with his scrotum containing two small testes and a lump and woke to find it containing two medium-sized 'falsies' that, in the years to come, many a girl was to fondle without ever guessing their true nature.

It was a month before he was shown the rat colony. But on his first visit he was feeling too ill from the combined radio- and chemotherapy to really appreciate what they would do for him. Six months later, however, he began visiting the colony weekly and marvelling at what he saw. There they were, the guardians and nurturers of his reproductive future. Ninety per cent of his testicular tissue had been removed during his operation, leaving just enough in his scrotum to support the plumbing that normally carried sperm from his testes to his penis. Even so, once all the potentially cancerous tissue had been cut out of what had been removed, there had been relatively little left. What there was had been in the deep-freeze for seven months before the decision was made to inject some of his precious cells into the testes of just ten male rats. And there they were, his sperm's nursemaids, running around their cages, living so that he might procreate. He found it difficult to believe that they contained part of him. But the doctors assured him his testicular tissue was still there. Moreover, it was alive and healthy and producing healthy sperm – which it wouldn't have been, had they left it inside his scrotum while he underwent his treatment.

On one of the man's visits to the colony, a rat was anaesthetized for him, had tiny electrodes placed in strategic places, and was made to ejaculate. Then, through a microscope, he was shown the sperm the rodent had produced. Once his eye was in, he had no difficulty spotting his sperm amongst the rat's. Rat sperm, he was told, had hook-shaped heads. His were the oval-headed sperm, looking like animated paddles. He watched in amazement as his seed jostled for position amongst the hook-headed hordes.

'What would happen if one of these rats had sex with another rat?' he asked, with visions of fathering a monstrous rat-man with a female rat. 'What would happen to my sperm?'

'They would probably live and swim around for a while,' the doctor replied, smiling the smile of someone who had answered this question many times before. 'They might even reach an egg, but fertilization is many, many times more likely by a rat sperm.

Even if, by some remote chance, one of your sperm did fertilize an egg, it wouldn't develop. The egg would just die.'

A year after the man's operation, the decision was made to kill the rats, remove their testes, graft them into his scrotum, and join their plumbing with his. As far as the medical team could tell, his therapy had worked and all traces of the cancer had been removed from his body. It was safe for him to have his balls back – or rather, the rats' balls.

He felt ill for a while, suffering the side-effects of the treatment designed to stop his body from rejecting the grafts. He empathized with what the rats would have gone through on his behalf while they were being prepared to receive his tissue, then later when they were killed so that he could have their testes. Then, three months later, there began a series of weekly visits to the clinic to test whether he was producing sperm and whether the plumbing had been joined up correctly.

The first few visits were abortive. Frightened that he might never ejaculate again, petrified that if he masturbated too hard his patchwork testes would disintegrate, and still a little sore, he was totally unable to excite himself. But after a wet dream had calmed some of his fears and gentle practice at home had restored some of his confidence, he eventually began to enjoy his visits. Shut in a room with films and magazines, he slowly began to rediscover the sexual fire of which he had been so proud before discovering his lump.

There were no sperm in his semen at first, then a few, then rather a lot – or at least, that's how it seemed to him when he was first given the chance to look down a microscope at what he had produced. Most of what he saw was rat sperm but there was no doubt there were many of his there as well. According to the doctor, it was only a hundredth of what he would have been producing before the operation, but it was enough to give him some chance of fathering children by intercourse.

He stroked the head of the tiny girl asleep by his side on the settee – and here was the proof that they had been right. He had met his current partner three years after he had been signed off by the medical profession and sent forth to procreate. Two years later, she had conceived their son – and surprisingly easily at that. It had taken only five months of unprotected sex.

During those five months he had gone back to see his counsellor for he needed advice again, unable to decide whether to tell his partner about his history. He thought she had a right to know that the scrotum she caressed was partly his and partly a rat's – but mainly sponge. He thought she also had a right to know that if it happened at all they could expect conception to take longer than average – perhaps much longer. But more than anything, he thought she had a right to know that every time he inseminated her, hordes of rat sperm as well as his own swam around inside her. His counsellor, however, while saying the decision was of course entirely his, advised him not to tell her. Certainly not before she had conceived and preferably never. 'Not everybody understands,' she warned. He never did tell his partner, and when she conceived, honesty no longer seemed necessary.

During his partner's pregnancy, new fears sent him scurrying back to his counsellor yet again. Surely it was surprising that she had conceived so easily? Surely it should have taken longer? His partner was a free spirit, attractive to men, and had had many partners during her life. Maybe the child wasn't his? Maybe, despite everything he had been through, he was infertile after all? His counsellor had done her best to reassure him and to urge him to be patient, and once again, her advice had been sound. The routine paternity test after his son's birth showed that the baby really was his – and four years later, so was his daughter. The rats had done a good job on his behalf. He had tried to repay them though, in the only way he knew how.

Shortly after his son had been born, the man had bought a female rat. The litter she produced was the beginning of the colony he had shown his daughter earlier that evening. At times, his partner was jealous of the hours he spent with his rats. But even without knowing his story, she tolerated them, not least because her children enjoyed watching them so much. Little did she realize the part they had played in his life – or how strong her grounds for jealousy really were.

Soon after his daughter had fallen asleep, the man's partner returned from her evening out. Tired and irritable, she put her daughter to bed and then showered and went to bed herself. Making his excuses, the man said it would be about an hour before he came too. He had things to do and he wanted to check the pregnant rat, to see if she had given birth yet. His partner grunted and pulled the sheets over her head.

The man went down to the rat-room in the cellar and closed the door behind him. He walked straight to the compartment with the pregnant female, then despite himself exclaimed with delight, 'Ah, my babies!' as he saw that she had given birth. They were too well hidden to count, but he stood watching with paternal pride as the mother licked and groomed her naked, pink young.

After a few minutes, he moved on to another section that he had isolated days ago. It housed another female, who came to greet him, nose twitching, eyes staring impassively. Maybe it was time, he thought. He slipped a glove made of rabbit fur onto his right hand and put it into the compartment, chasing the female around and stroking her back when he could. After a while, she crouched on the floor, immobile, in the position that, had he been a male rat, would have been an invitation to sex.

The man smiled and walked over to a corner of the room to take a beaker, a pipette and a tiny cork bung out of the cupboard. Dropping his trousers and pants, he began to masturbate. Just three minutes later he was ejaculating into the beaker. Filling the pipette with his semen, he walked back to the frustrated female. With a practised grip, he removed her from her compartment, inserted the pipette into her vagina, and injected his semen. Then he pushed the small cork bung up inside her to help retain the fluid before putting her back into her compartment.

The first female he had bought after his son had been born had *not* already been pregnant when he bought her. He had fertilized her in just this way using the rat sperm in his own semen. All the rats in his colony were descendants of the rats who had kept his own testes alive, ten years ago, so that he could reproduce. The least he could do, he had decided, was the same for them.

A Life without Testes

This is the second man we have met in this book who has had to come to terms with having no testes. The first man, in a wheelchair in Scene 4, lost his testes through accident. This man lost his through cancer. Both lost their ability to manufacture sperm and to be able to procreate via intercourse.

Accidents are sudden, leaving no room for forward planning. We can only assume that the cripple in Scene 4 lost his genitals at the scene of the accident. Otherwise, the surgeons who repaired him might well have been able to save and cryopreserve (freeze) some of his testicular tissue while decisions were made over what to do. In which case he, too, could have used the rats to make sperm for him. There would have been no need for haploid gametes to be manufactured by enforced meiosis as described in Chapter 4.

The rat man in this scene had the relative luxury of forward planning and counselling in making his decision. He had, in fact, many options and we already know that he took the precaution of banking sperm and cryopreserving stem cells as well as taking the rat option. His decision undoubtedly betrays something about his character but it also betrays something about the time in which he was living. Evidently, it was a time in which some people still held precious the option of making babies via intercourse – even among those who could afford to do otherwise. The year, then, is probably still around 2050, contemporary with Scene 4.

There is much to say in later chapters about the change in people's attitudes that will probably be occurring around 2050. For the moment, though, let's concentrate on the option outlined in the scene – because with more and more men destined to lose their testes in the future, an increasing number may have to contemplate placing their testes in the hands, so to speak, of rats.

Sperm Falling, Cancer Rising

In 1992, researchers at the University of Copenhagen sounded an alarm that was heard around the world. After analysing data from studies going back to the 1930s, they concluded that there had been a dramatic decline in sperm counts in the last half of the twentieth century. Since then, research at other European centres, such as Paris and Edinburgh, has also found examples of falling sperm counts since the 1960s. And although most studies have looked at the number of sperm men ejaculate during masturbation, a study at the University of Manchester in the UK found a decline in the numbers ejaculated during intercourse as well.

Such research has triggered nightmare scenarios for the future, such as that portrayed by P. D. James in *The Children of Men*.

She describes a situation in which the quality of sperm declines steadily in the early years of the twenty-first century until hardly anyone can reproduce in the normal way. At the same time, the countryside becomes virtually emptied of animals as natural populations crash. The first signs of catastrophe were noted in the 1990s, but few people could believe that simple environmental pollution could have such a diabolical effect. And so nothing was done – until it was too late.

Part of the reason that most scientists in the 1990s were unconvinced that the future was so bleak was that not all research on sperm counts showed a decline. Even in Europe, studies in Finland and France (Toulouse) did not, and in America they certainly didn't. In fact, American men's sperm seemed to be doing rather well compared with their European counterparts. At the University of Washington in Seattle sperm counts of students who donated sperm for research between 1972 and 1993 showed no decline. And sperm counts of men in Minnesota, New York City and California who banked sperm between 1970 and 1994 before they had a vasectomy actually showed a significant increase in New York and Minnesota and a slight increase in California.

These conflicting conclusions have caused the sperm count studies to be analysed and reanalysed – then supported or criticized – from almost every statistical and methodological angle. As a result, for every scientist who claimed there was cause for alarm, there was another of equal eminence who would say that there was nothing to worry about.

In January 1997, however, Finnish scientists published the most convincing evidence yet that sperm production by men in industrialized countries is declining. It was more convincing than the studies that went before because it side-stepped the notorious unreliability of counting sperm in semen samples. Instead the researchers at the University of Helsinki examined tissue from the testes of hundreds of middle-aged Finnish men who died suddenly in either 1981 or 1991. Among the men who died in 1981, 56 per cent had normal, healthy sperm production. By 1991, however, this figure had dropped dramatically to 27 per cent. The average weight of the men's testes decreased over the decade, while the proportion of useless, fibrous tissue increased.

So sperm production may really be falling. Even more convincing, as indicated in the scene, is the evidence that since the 1930s there has been an increase in several problems with male reproductive health. These include cancer, malformations of the penis and undescended testicles. The incidence of testicular cancer is increasing every year, and in Denmark now affects 1 per cent of young men.

Even if male reproductive health *is* in decline, as the various studies suggest, scientists are still some way from identifying a clear culprit. Many believe that chemicals from women's urine in sewage are the most likely contenders. Others think that industrial pollution is more likely. Wherever the problem originates, though, the general opinion is that an abnormally large amount of female hormone – oestrogen or oestrogen-mimicking chemicals – in the environment is most likely to be the main cause. In recent years, headlines such as 'Men unmanned by a sea of oestrogen' have abounded.

Certainly, the list of chemicals alleged to mimic or interfere with sex hormones is now vast. And for people who hate long scientific words and acronyms it reads like a nightmare in more ways than one. It includes ubiquitous industrial chemicals and products of petrol combustion such as polycyclic aromatic hydrocarbons (PAHs), polychlorinated biphenyls (PCBs) and dioxins. There are phthalates that are added as plasticizers in plastics and used as ingredients in paints, inks and adhesives. Alkyl phenolic substances (such as octyl and nonyl phenol), which are breakdown products of alkylphenol polyethoxylates (APEs) used as surfactants in industrial detergents and also found in paints, herbicides and some plastics, have also been suggested. And so, too, have organochlorine pesticides such as DDT, aldrin and dieldrin.

An Anglo-Danish research team attempted to link all the evidence and ideas together by arguing that man-made oestrogenic chemicals from the environment might be damaging male foetuses in the womb. Sex hormones such as oestrogen influence the development of the reproductive system in the developing foetus, including the nurse cells that control future sperm production. And since sex hormones can also stimulate the proliferation of cancer cells, in theory environmental chemicals that mimic them could be responsible for the rise in testicular cancer.

One problem with attempts to blame human health problems on oestrogenic chemicals is that a mother's bloodstream is awash with oestrogen during pregnancy. Every foetus is exposed to the chemical naturally. Equally, many foods contain chemicals that are oestrogen mimics. The Japanese diet, for example, is extraordinarily rich in oestrogen-mimics produced by plants such as soya beans. Yet there is no sign of increased infertility in Japan, and the nation's cancer rates are among the lowest in the world. Indeed, nutritionists suspect that a diet high in plant oestrogens may actively lower a woman's risk of developing breast cancer.

It's quite likely, though, that humans and other mammals have evolved to cope with exposure to ancient oestrogens and oestrogen-mimics such as those in a mother's bloodstream and in food. Man-made chemicals that mimic hormones, though, are a new phenomenon to which natural selection has not yet had time to respond. Certainly, the ability of synthetic chemicals such as DDT, phthalates and APEs to accumulate in fats in the body sets them apart from natural plant oestrogens and may make them more dangerous. Algae and fish can take up these chemicals so that trace amounts in rivers, lakes and the sea become multiplied thousands of times in food. The chemicals can then be stored in human fat; if this fat is broken down as often happens early in pregnancy, concentrated pollutants may flood the system.

Medical researchers have already glimpsed what might happen later in life to foetuses exposed to an excess of synthetic hormones during pregnancy, thanks to a famous 'medical mistake'. Between the 1940s and the late 1970s, doctors prescribed a synthetic oestrogen drug, DES or diethylstilbestrol, to an estimated 2.3 million pregnant women worldwide in the belief that it would prevent miscarriage. The adult children of the women given DES are now acknowledged to be at greater risk of developing various cancers of the reproductive system. They also suffer a higher incidence of infertility and other reproductive problems. This could have been a direct effect of the drug. Alternatively, it could have been due to women being prevented from miscarrying foetuses that their bodies might normally have detected harboured problems.

It is currently very difficult to assess whether environmental oestrogens might really be responsible for a fall in sperm counts or rising rates of testicular cancer. Most governments still feel that a ban on suspected chemicals would be premature. Even the fact that certain substances *could* damage reproductive health doesn't necessarily mean that the chemicals people are currently exposed to in their everyday lives are actually doing so. Establishing such a link scientifically is no easy task, and most of the evidence accumulated so far is dogged by uncertainty.

Future men, like the main character in the scene, may or may not find that their and their sons' reproductive health is less secure than now. And if they do, they may well welcome the opportunity provided by one of the most bizarre of recent technological developments – surrogate testes.

Surrogate Testes

In May 1996, a team of scientists from the Universities of Pennsylvania and Texas announced in the respected scientific journal *Nature* that they had succeeded in 'persuading' male mice to produce rat sperm. In the same paper, they speculated that this was just one step on the way to being able to incubate the sperm from a whole range of mammals within the testes of other species.

What the researchers had done in their original work was to inject cells from rat testes into the testes of special mice. The mice not only accepted the foreign cells but also nurtured and matured them into fully formed rat sperm. The cells that were injected were the stem cells described earlier. In the intact testes, these stem cells are associated with specialized nurse cells that nurture and support them, as we have also seen. In the experiment, the researchers injected the mice with not only stem cells from the rat but also nurse cells, thinking that the latter would be needed for the successful production of rat sperm in the mouse. What they found, however, was that the rat's sperm were nursed and supported not by their own nurse cells but by those of the mouse.

This experiment worked because the strains of mice used to receive the rat cells were special. They were immuno-deficient. In other words, their immune system was so weak that they did not kill the foreign rat cells. These mice are perfect recipients for foreign tissue and maybe the scene we have just witnessed should have been called 'One man and his mice' instead of 'One man and his rats'. However, even before their original *Nature* paper had been published, the same scientific team had succeeded in injecting testicular cells from one strain of rat to another. If transfer of testicular cells from different species into a reservoir species is to escalate in the future, it may be that the larger size of the rat testis will make rats more suitable recipients than mice. The first application for a licence to transfer human stem cells into another species, though – which was made public in 1998 – named the mouse as the surrogate species.

The reason for injecting stem cells into the host's testes rather than grafting chunks of testicular tissue is to solve the problem of 'plumbing'. Stem cells line the walls of a large number of tiny, convoluted seminiferous tubules in the intact testis. The problem of transplanting chunks of testicular tissue composed more or less entirely of these tiny tubules is that the sperm can never find their way out to enter main tubes and be ejaculated. Surgically, it would be impossible to join up the tiny tubules of the host and donor in any effective way. By injecting stem cells directly into the host's tubules and allowing them to attach to the tubule walls, the problem is solved. The host's tube and tubule system is still intact. Any sperm produced will automatically find their way into the host's main tubes.

Plumbing is less of a problem if a whole testis can be transplanted from donor to host, because their large tubes can be connected together, just like joining blood vessels during heart transplant surgery. This is why, in the scene, it was just cells from the man's testes that were injected into the rat but whole rat testes that were grafted back into the man's artificial scrotum. It was then 'simply' a matter of joining the tubes from the rat's testes to those of the man's that were spared during the initial surgery. That way, the rat's tubule system could be used to carry both rat and human sperm to the man's main sperm tubes – the two epididymides and vasa deferentia – and hence eventually to his penis for ejaculation. One problem with this technique, though, is that it means the man ejaculates rat sperm as well as his own.

In fact, it may not have been necessary for the man to endure the production of rat sperm during his treatment. It would have been possible, before the man's stem cells were ever injected into the rats, to expose the rats to chemicals – busalphan was used in the original experiment – which would destroy the rats' ability to produce their own sperm. This would leave the rat testes free for the production of only human sperm when the man's stem cells were later injected. However, the danger that such devastating chemical exposure might subsequently have deleterious, perhaps genetic, consequences for the man's sperm might be considerable. Unpleasant though the thought may be, allowing the production of rat sperm to continue would probably be the safer option.

The prospect of receiving a transplant of rat testes may seem repulsive to a man. And the prospect of her partner possessing rat testes and inseminating rat sperm along with his own may be equally repulsive to a woman. But in reality it is no different from receiving any organ from a non-human species.

Such 'xenotransplantation' hasn't happened yet, but scientists are actively working on the technique. In 1992, a man suffering from liver damage due to hepatitis B arrived at Duke University Medical Center for treatment. Doctors linked him up to a succession of five pig livers while they waited for a suitable human donor. The man survived and recovered. In 1995, a man suffering from AIDS received a bone-marrow transplant from a baboon. His health recovered, though probably not due to the transplant, which eventually failed.

Perhaps surprisingly, a 1998 poll revealed that more than three-quarters of the US public would consider a xenotransplant for a loved one if the organ or tissue were not available from a human. Respondents preferred baboons or chimpanzees as donors rather than pigs – they were not asked how they felt about testes from rats!

There are still a number of obstacles to overcome before xenotransplants can become a powerful medical tool, the main problems being immunological and the spread of viruses from non-humans to humans. One approach being pursued is the use of genetically engineered and virus-free pigs as a source of transplant hearts and livers. The transplant of rat testes into men's scrotal sacs is in principle no different. The main danger is that rat viruses will spread to humans via either the transplants or the sperm. On top of that is the danger that development in a rat testis might change the sperm DNA in some way, causing the same sorts of developmental problems that we have already discussed in relation to ICSI and ROSI. But as with these latter two techniques, we can only finally decide whether the dangers are real or imaginary by going ahead with the technique and then monitoring the children so produced.

In the scene, it was rather optimistically assumed that the man, with his bevy of transplanted rat testes, would produce enough sperm to be able to fertilize his partner via intercourse. In the original experiments, the number of rat sperm produced by the mouse testes was rather small and no attempt was made to determine whether those sperm would fertilize a rat – or to transplant the mouse testes back into a rat. The first step to be taken by the laboratory in Melbourne, Australia, which has applied for a licence to extend the work to humans, will be to see if rat eggs can be fertilized by the rat sperm produced by a mouse testis. They will undoubtedly do so either by standard IVF, or more likely by ICSI, rather than transplanting the mouse testes back into a rat and encouraging intercourse. In truth, this is likely to be the most common use of the technique for humans also. Whether there will ever be a demand to have the rat testes transplanted back into the man so that he can reproduce 'normally' – as in the scene – remains to be seen. Much depends on the relative timing of the developments described in this scene (Scene 5) and those described in Scenes 6–8.

Although the prospect of being inseminated with rat sperm

will seem repugnant to most women, it is not such a major departure as it might seem. Every time any woman is inseminated she invariably receives a whole mixture of foreign cells – bacteria, viruses and protozoa – either directly from the man's penis or mixed in with his semen. It is partly for this reason that, immediately she is inseminated, the woman's body unleashes hordes of white blood cells in her womb and cervix. Their job is to find and kill foreign bodies such as these invading cells – including, as we have seen, a large proportion of the human sperm that she receives from her partner. Rat sperm would simply be treated like any other foreign cell and the vast majority would be killed by her white blood cells within hours of being ejaculated.

We should not forget, either, that although most women undoubtedly find the prospect of being inseminated with the sperm from another animal unpleasant, not all do. A tiny number of women (less than 1 per cent according to the US survey by Kinsey in the 1940s) actually go out of their way at least once in their lives to have sex with other animals – mainly dogs – and so receive a full ejaculate of another species' sperm.

Surrogate Ovaries

Although this chapter is concerned primarily with the treatment of a particular form of *male* infertility, we should note that at least part of the technology is also being developed for women. Instead of harvesting a woman's eggs, bits of her egg-bearing ovarian tissue could be removed and cryopreserved. Like egg freezing, this procedure would preserve fertility for women who know that they are about to lose their ovaries. It could be used on females who are far too young to produce mature eggs – girls who are undergoing radiation treatments, for example. In theory, the ovarian tissue could eventually be placed back in the body and lead to successful pregnancies – as has been done in sheep but not yet in humans.

The latest research uses mice as 'incubators'. As with stem-cell transplant into testes, immature eggs from sheep have been transplanted into immuno-deficient mice that do not reject foreign tissue. There, in the mouse ovaries, the sheep eggs survived and matured without harm. The technique could obviously help zoos to conserve rare species. But there is no reason in theory why human eggs should not be matured in the same way (Scene 7).

Fantastic Breeding Potential

The main use of the rat option in the future is likely to be that depicted in the scene – a means of resuming production of sperm after extensive radiotherapy or chemotherapy. Such treatment either kills stem cells or raises the risk of such horrendous mutations that the man is essentially incapable of producing children thereafter. It is already possible to cryopreserve stem cells, and they can be left indefinitely in storage until a man is ready to use them. As in the scene, they would then be injected into the testis of a rat, or rather several rats. Then either the rat testes – or most often just the sperm they produced – would be used as the man required.

There is a second possible use in the treatment of infertility. Failure of a testis to produce sperm may be the result of faulty nurse cells rather than faulty stem cells. Yet, as we have seen, transplanted stem cells utilize the *host's* nurse cells rather than their own. Stem cells that in the man fail to produce sperm may do so quite successfully in another testis. In this sense, then, surrogate testes may be an alternative to ROSI in the treatment of this particular category of infertile men.

Although the main use of the rat option is likely to be in the treatment of *human* testicular cancer or infertility – all individuals being 'valuable' – a secondary use could be in the treatment of valuable individual males (race-horses, pedigree dogs, etc.) of other animals. From time to time such individuals need treatment for diseases, including testicular cancer, that would involve destroying their ability to produce sperm. In such cases the rat option would be an ideal way of allowing them to continue to produce sperm after their treatment. Stem cells could also be collected from the males of endangered species, cryopreserved, then injected into rat testes for the production of sperm if it ever became necessary. Of course, for these other species there would certainly be no need to transplant the rat testes back into the original male. The sperm could be collected from the rats themselves – by electro-ejaculation as described in the scene – then fertilization and reproduction of the main species could be achieved by ICSI and embryo transfer.

One of the problems with the use of rats to nurture the stem cells of men or other species is that rats live for only a few years. Once stem cells have been injected, they will be lost when the rat dies. Of course, cryopreservation of enough cells from the original donor would mean that many generations of rats could be injected, guaranteeing production of the donor's sperm for as long as they were likely to be required. Every time the donor's sperm were needed, some of his stem cells could be removed from storage, injected into a new male rat, and the sperm eventually produced then used.

The only drawback to this system is that it requires some

forward planning for there will be around a three-month delay between seeing a need for the donor's sperm and the sperm being available from the host rat. There is an alternative, however, and that is to use a long-lived animal as the nurturer of another species' sperm cells. Then, sperm will be just an ejaculation away for the duration of the host's long life. Several long-lived species seem possibilities as hosts – elephants come to mind – but none of them are terribly convenient as testis nurses. In the case of elephants, one problem is that the testes are tucked safely away deep inside the body, not dangling dangerously albeit conveniently as in humans.

By far the most convenient host for other species would be man himself. He is long-lived and can ejaculate to order without the need for electro-ejaculation. Most men, though, are unlikely to volunteer to nurse the sperm cells of another animal, no matter how precious the individual nor how endangered the species.

Nevertheless, every technological and social change spawns minority practitioners. The man in the scene found his own reasons for fertilizing female rats with his semen – an act of gratitude to the male rats for giving him the two children that were now so precious to him. According to Kinsey's survey of American men in the 1940s, nearly one in five of those raised on farms claimed that at least once in their lives they had copulated with livestock. Given this, how many sheep-farmers would consider having their testes injected with stem cells from a favourite ram so that they could produce lambs by copulating with ewes themselves? Or how many dog-breeders would do the same with their female dogs? Or how many women would have stem cells from a favoured, perhaps dead, human partner injected into the testes of their pet dog, giving themselves the option of having further children sired by their deceased partner by having intercourse with their dog?

Such fantasies apart, the rat option – probably in conjunction with ICSI rather than intercourse – does offer one avenue to continued fertility for men who are suddenly faced with a life without testes. In the future, though, there will be other options, as the remainder of this book will gradually explore.

6

Cloning

Resurrection

'Cuddle me!' said the two-year-old, arms open wide as she launched herself at her mother.

The woman sat firm and resolute for all of two seconds. She had just reprimanded her daughter for jumping up and down on the settee – the same settee she was sitting on, trying to drink a cup of hot coffee.

'Stop it, Phee,' she had shouted as a slop of brown liquid projected itself out of her cup and over her freshly cleaned skirt. For three more jumps, her daughter had defied her, her long golden hair flouncing around her shoulders with each leap. 'Stop it,' the woman had repeated, more severely this time. 'You *know* you're not supposed to jump on the furniture.'

A third reprimand, shouted sternly, had been necessary before the little girl had finally stopped. Then her face crumpled and she began to cry. Not a real cry – just enough to make her mother feel guilty and finally to laugh at the expression on her daughter's face. It was then that the girl had asked for a cuddle. The coffee was nearly spilled for a second time as the two began their act of affection.

Awkwardly – for nothing was easy with a two-year-old wrapped around her neck – the mother put her cup down on the floor, then put her arms around her daughter's waist, holding her close and kissing her on the cheek. These were the moments that made life worthwhile.

Three hours later, bathed, powdered and dressed in pyjamas, the rounded peaches-and-cream face was all that could be seen as the girl snuggled into her bed.

'Just one nursery rhyme tonight, darling,' said the mother. 'It's late, which one do you want?'

'Err....' The little girl thought for a moment. 'Err ... Dack and Dill,' she suddenly said, emphatically.

'OK.' The mother searched for the place in the book. Before beginning to read, she glanced up at her daughter's smiling face. It was another of those moments that suddenly hit her out of nowhere. The face gazing at her so expectantly – almost disembodied, framed as it was by the sheet and bedspread tucked under her chin – was so unnervingly like her dead father, it could almost have been a miniature of his.

Recovering her composure, the woman read the nursery rhyme slowly, then painstakingly and patiently went through each of the pictures that illustrated it. The little girl particularly liked the drawing of Jack and Jill coming down the hill, but wondered why Jack wasn't wearing a crown. In fact, the mother had to read the rhyme and show the pictures several times before she was eventually allowed out of the room.

First, she tidied the kitchen and lounge, returning the toys that had been scattered in a thin layer over the floor to their rightful boxes and positions, fretting over missing bricks and tiny pieces of games. Then, eventually, she sank back on the settee and picked up a magazine. Bliss.

She still lived in the house that she had shared with her partner before he died seven years ago. In truth, it was too big for just her and her little girl, but she was reluctant to sell. Apart from her daughter, the house was her only real link to the man she had loved so much. There were times she could actually imagine he was still present, and she gained reassurance from that.

He would have been so proud of his daughter – but he had never seen her. The nearest he had been was watching her ultrasound image when she was just fourteen weeks old in her mother's womb and later, as he lay dying in his hospital bed, feeling her kick against his hand as she wriggled in her mother's stomach. It was almost the last sensation he ever felt.

'Damn paper-boys,' said the woman two hours later as she switched off lights and closed doors in preparation for going to bed. Out of the front window, she noticed that the gate to their drive was open. It was a double wrought-iron gate that she closed religiously every time she drove the car in or out of the drive. In the three years since she'd had the gate fitted, keeping it closed had been one of her obsessions. Every paper-boy or postman who had ever delivered to their house had been chastized at one time or other for not closing the gate after them.

She couldn't go to bed and leave it open, so she put on her shoes and coat to walk out and close it, taking care to put the door on the catch so as not to lock herself out. It was a steep slope from their front door down to the road and a slight frost was forming. She walked gingerly to avoid slipping. The iron gate was cold to her hand as she closed it.

The next morning, mother and daughter had their usual leisurely breakfast. There was a storm outside – wind and rain – and the little girl was fascinated by the way the tree outside the kitchen window was swaying.

'Eat your toast, Phee. Be a good girl for Mummy. We're going out to the shops soon.'

The little girl pulled a face and pushed away her plate, her mouth and cheeks streaked with a mixture of butter and chocolate spread. She shook her head and picked up her drink instead. She never did as she was told – not first time, anyway. But she usually disobeyed so engagingly that her mother rarely lost her temper.

'OK. If you don't want any more, let's go and clean you up and get ourselves dressed.'

As the woman went to lift Phee down from the tall chair, the little girl threw her arms round her mother's neck. The woman knew – she just knew – that she now had butter and chocolate all over her hair.

It was another hour before mother and daughter stepped out of their front door. The little girl was dressed in a short red plastic mackintosh, the mother in a long flowing coat that she hadn't bothered to button. It was only a few steps to the car, parked at the top of the drive. As the mother closed the front door, a sudden gust of wind made it slam shut, strong enough to take their breath away. The little girl laughed at the sensation.

'Come on, Phee, get in quick,' said the mother as she opened the car door. 'I'm getting wet.'

The girl didn't hear her mother's plea, the words drowned by the noise made by a juggernaut thundering past the front gate, followed by two cars. In any case, she was far too excited by the wind and rain to climb meekly into the car.

'Now, don't run off,' said the mother, recognizing her daughter's body language. It was a frequent game – that's why the gates were there – and one that the mother hated. There was no problem. She could catch her long before she reached the gates – it was just the memories.

'Phee, get in the car. Please, darling. No, don't run off. Phee, please. No, Phee don't you dare. Come HERE.'

But her daughter took no notice. 'Weeee,' she shouted excitedly as she began to run down the slope. 'Look, Mummy. Dack and Dill.' She ran fast, waiting for the moment that her mother would grab hold of her. She loved the game – oblivious to the danger on the other side of the gate.

As the mother let go of the car door to give chase, a gust of wind simultaneously whipped her long light coat up around her and slammed the car door on to it. When she tried to run, she was held fast. It took only seconds for her to open the door again, but it was long enough. By the time she set off, her daughter was almost halfway to the gate, left ajar by the early-morning postman.

'Phee, stop,' she screamed as she gave chase, realizing she couldn't catch her. 'Phee, please . . . Phee! PHEE! . . . PHOENIX! STOP.'

Still yards away from catching up, she saw the little red figure run straight through the gate and on to the road. Brakes screeched as cars travelling in both directions tried to stop. Suddenly, her daughter was hidden from view by a van that skidded to a halt in front of her gates. There was more screeching followed by the sound of crunching metal and breaking glass as one car ran into the back of another. As she scrambled round the front of the van blocking her way, her stomach felt as if it was trying to force its way out of her mouth. Not again, she thought. It couldn't happen to me – to her – not again.

When she rounded the van, she almost couldn't believe her eyes. Her daughter was standing in the middle of the road, screaming with fear – but unharmed. She grabbed her up into her arms and held her tight. Oblivious to the shouts and curses of drivers relieved they hadn't killed the little girl but furious at the accident she had caused, the woman took her straight into the house. The pair sat on the settee, still dressed in their coats, wet from rain, and clinging to each other as if they would never part, both crying from fear and relief.

The woman's mind was a maelstrom of thoughts. She would sell the house – move somewhere safe. The gate had been supposed to stop it happening again. Damned postmen. Damned paper-boys. Maybe she shouldn't have called her Phoenix. At least, not the second time. But it had been her partner's choice. He wasn't to know how macabrely appropriate it would be.

Silly thoughts entered her head and refused to leave. Maybe Phoenix was destined never to survive beyond the age of three.

Maybe death was in her genes. But she would do it again. She'd had no hesitation over cloning the first time – and she wouldn't have hesitated today. Four years ago, she made the decision in the ambulance – when the paramedic stood over the crumpled body of the first Phoenix and pronounced her dead. She had begged them to make sure some cells would stay alive long enough for cloning to be possible.

She'd had the counselling and heard the arguments. Let go of the past. Let her die. Start again. But she couldn't. She loved Phoenix so much, and she'd loved Phoenix's father so much – she just couldn't let her die. Not if there was an alternative – as there was. And she'd never regretted it. OK, there had been differences between Phoenix-one and Phoenix-two. But they were trivial. To all intents and purposes, they were the same beautiful, loving child – she was just taking a little longer over growing up, that's all. She would have been seven; instead she was nearly three.

The woman held Phoenix close, hugging her as tightly as she could. 'Cuddle Mummy, darling,' she sobbed.

Cloning: Fears, Fantasies and the Future

Set roughly in 2050, this scene should have been a surprise. Phoenix was a clone – but *not* a nightmare. Instead she was a pretty and lovable individual, full of character and life, and her mother adored her.

Of all the novel ways of reproducing in the future, cloning is the one that has most captured the public imagination. Writers and general public alike are all aware of the bizarre prospect of people being cloned – having genetically identical counterparts –

and many horrific scenarios have been described. This scene, though, was not one of them.

Nightmares about cloning are not new. In *Brave New World*, for example, published in 1932, the Central London Hatchery and Conditioning Centre immersed human eggs in a 'warm bouillon of free-swimming spermatozoa'. Then 'bokanovsk-ification' made the resulting embryos bud into hundreds of identical cloned Gamma, Delta and Epsilon human beings. More typically, people in the past have imagined clones of themselves being produced by some form of budding or bodily splitting – a sort of sideways reproduction – almost as if the image in a mirror magically comes to life.

Such visions are very different from the glimpse of cloning that recent technical developments have provided. And as this scene has suggested, for some people cloning could be a boon, not a bane.

Cloning – the Technology

The scientific breakthrough that has brought human cloning so near does not work by budding or splitting. It *will not* be possible for a person to decide that they would like somebody identical to themselves – same age, same appearance and same character-istics – suddenly to exist. If, by the time a person is adult, they do not already have a clone-brother or clone-sister, then it is too late. The best an *adult* will be able to do is to have a clone-son or clone-daughter – genetically identical to himself or herself but separated in age by a generation – as will be illustrated in Scene

10. The reason is that even clones begin life as embryos and need a surrogate mother to carry and give birth to them.

The first animal to be cloned successfully from an adult animal was a sheep. On 27 February 1997, it was announced in the journal *Nature* that a team of scientists from Scotland had successfully produced the first individual – named 'Dolly' – to be cloned from cells taken from an adult. The method used was as follows.

The sheep cloned – the 'clone-mother' – was a six-year-old ewe of a breed known as Finn Dorset. She was in the last trimester of pregnancy when cells were taken from her mammary glands. These udder cells were then cultured in an artificial serum that nourished them, allowing them to live and multiply. Each cell in the culture carried in its nucleus all the genes possessed by the female clone-mother. These contained all the instructions necessary to produce another individual genetically identical to the clone-mother. What each nucleus did not have was a cell around it that would allow, or even trigger, its genes to begin giving those instructions. Udder cells are udder cells – they cannot suddenly begin to turn into all the different types of cells that are needed to make a whole sheep. Only one cell has that capacity – the unfertilized egg, known as an 'oocyte'. What these udder nuclei with their genetic instructions needed, therefore, was to be placed inside an oocyte which they could then instruct to develop into a whole sheep.

The oocytes used were taken from a breed of sheep – Scottish Blackface – that was different from the clone-mother's breed. This was done deliberately so that the experimenters could check that any sheep produced by their experiments was visibly a clone of the female who donated the udder nuclei, not offspring of the female who donated the oocyte. The Blackface egg donors were injected with gonadotrophin-releasing hormone to make them release oocytes from their ovaries. Then, twenty-eight to thirty-three hours after injection, the oocytes were collected. The only problem, of course, was that each of these oocytes already contained its own nucleus with instructions – or rather half the instructions – to make a Scottish Blackface sheep. However, technology is now sophisticated enough for the experimenters to be able to remove the Blackface nucleus from each oocyte and replace it with a nucleus from the Finn Dorset clone-mother.

Normally, an oocyte is stimulated to begin developing into an embryo by fertilization – penetration by a sperm. However, eggs can be tricked into reacting as if they have been fertilized in various ways. Adding calcium is one way, electric impulses another. In the cloning experiment that produced Dolly, electrical impulses were used both to fuse the new nucleus with the oocyte and to trigger development. Only thirty-four to thirty-six hours after the Blackface ewes had been injected with hormone, oocytes now complete with replacement nuclei from the clone-mother were being stimulated to develop. As they did so, the genetic instructions they followed were those from the six-year-old Finn Dorset, every embryo genetically identical to its clone-mother.

At first, the developing embryos were injected into oviducts in live sheep, the oviducts being tied at both ends to hold the embryo in place. Then, after six days, those embryos that still survived and seemed to be growing healthily were transferred into the wombs of surrogate Blackface mothers and allowed to develop to full term via 'normal' pregnancy.

In this pioneering experiment, the process was very unreliable. Dolly was the only lamb born from 277 fusions of oocytes with nuclei from udder cells. The team had greater success (three lambs born) if they used cells taken from a 26-day-old foetus and even greater success (four lambs born) if they used cells from a nine-day-old embryo. The younger the cell, the more easily it cloned. There were also a number of miscarriages and foetuses with abnormalities. However, this was only the first experiment – and there can be little doubt that over the next few years a reliable process for cloning from adult cells will be achieved.

There are a number of scientific questions that can be tackled immediately using cloning techniques. The most important concerns the effects of cell differentiation on the genes they contain. In this context, the very fact that Dolly exists is by itself enormously instructive. Until then, it was thought that genes were irreversibly switched off during development. Every cell in the body contains an identical – and complete – set of genes. But in a liver cell, for example, most of the genes have been 'switched off' so that only the genes that instruct the liver how to work are active. Similarly, a muscle cell only has genes switched on that give instructions to muscles. All the other genes are switched off and cannot be reactivated – or so it was thought.

On this theory, though, cloning from an adult cell would never have been possible, because depending on the tissue it was taken from, the nucleus would only give a few instructions – like how to be a liver. It couldn't give the full range of instructions necessary to grow a whole individual. This was why it was always thought that only embryonic cells could be used for cloning, as long as they were collected before any genes were switched off. And this is still probably why cloning has proved to be so much easier with embryonic and foetal cells. Dolly's existence, however, proves that the switched-off genes in adult cells can be switched back on. All that was necessary, it turned out, was to starve the cells, arrest them at a very precise stage in the cycle of cell division, and transfer them into an oocyte at that stage. Then – so it seems – all genes regain their expression.

Technically, anything that can be done with sheep can also be done with humans. Cloning has already been achieved with monkeys, though admittedly using the 'easy' route of taking cells from embryos so young that they consisted of only eight or sixteen cells. The question is, how long will it be before scientists clone humans?

The first – though probably abortive – scheme to open a human cloning clinic for the treatment of infertility was floated in Chicago in 1998. As a treatment for infertility, cloning would be infallible. Everybody has cells – all they need is an oocyte donor and surrogate mother and they can have their own genetic child. The only difference between reproduction by cloning and normal (gametic) reproduction is that the parent will be 100 per cent, not 50 per cent, related to their child, because a body cell contains all a person's genes whereas an egg or sperm cell contains only half. Cloning really would mark an end to both male and female infertility. Everybody could reproduce.

Cloning in Nature

In a biological sense, there is nothing new or scary about cloning. Spectacular and beautiful sights, ranging from swathes of vegetation to masses of coral, are often clones, colonies of genetically identical individuals each budded off from some other identical individual. In many such colonies of plants and animals, the cloned individuals stay attached, sharing a circulatory, nervous and/or root system. In others, though, they do not. In the familiar *Hydra* of school textbooks, the budded clone breaks free and takes on a free-living and independent existence. In yet other organisms, cloning is an adaptation for survival as well as for reproduction. Chop a starfish into tiny bits and each fragment will grow into a whole new being, each genetically identical to its mutilated clone-parent.

As far as humans are concerned, out of a World Population of about 6 billion people, roughly 48 million are clones – we call them identical twins. Protesters should be careful, therefore, when they condemn cloning as an assault on human individuality or dignity – there are many that could be offended.

There are, of course, two main types of twins, fraternal and identical, but only identical twins are clones, originating from a single zygote. Early on in the mother's pregnancy, the zygote's cell mass divides into two parts which develop into separate individuals – always of the same sex and with a more or less identical genetic makeup. Identical triplets, quadruplets and quintuplets have also occurred naturally – multiple clones arising from the zygote dividing into more than two parts – but are very rare. Identical twins aren't always identical genetically because mutations can occur after the two cell masses have separated. Recent studies of identical twins in which one has schizophrenia, for example, have found cases in which the critical DNA sequences differ between the two.

In the scene, Phoenix-one and Phoenix-two were not the same person. They were more like identical twins – it was just that they were born four years apart rather than four minutes apart and one of them had died. Clones will not look and behave identically, just because they are genetically identical – any more than identical twins do now. Clones will, of course, be very similar, but they won't be identical. There are two main reasons, one obvious and the other less so.

The less obvious reason is that chance mutations while the clone cells are being cultured – or during early embryonic development – could produce *genetic* differences between two clones, as we have just seen with schizophrenia in identical twins. On top of this, the genes resident in the nucleus of a fertilized egg do not control *absolutely all* of the individual's development. They control most of it, but the contents of the egg outside of the nucleus – the cytoplasm, or in this case, the ooplasm – probably also has some influence. In two or more cloned eggs, even if the different sets of genes are identical, the contents of the host oocytes will not be. As a result, no two individual clones will be absolutely identical. In fact, clones are less likely to be identical than identical twins, the clone-nuclei of which *do* share the contents of the same oocyte.

The more obvious reason is that upbringing and experience interact with genetic instructions to make a real difference to how people develop and behave. Clones raised together would be very similar – as are identical twins – but they would not be identical, because their experiences would not be identical. Slight differences in diet and diseases could have a big influence on appearance and behaviour. Stature and body symmetry, and hence health and behaviour, for example, are all influenced by diet and disease during development. Chance differences in experience can also make huge differences to appearance and behaviour. Suppose one clone but not the other just happened one day to be chased, bitten and licked on the mouth by a dog. At the very least, that person might thereafter be phobic about dogs whereas its clone may not. At worst, the victim may unknowingly catch toxacariosis from the dog's saliva and develop sight problems and occasional epilepsy, with all the implications those would have for future psychological development. Its clone, however, without such an experience and without such an infection, would have no such physical or psychological charac-teristics. This is obviously an extreme example, but even minor differences in experiences can lead to recognizable differences between people who are genetically identical.

Such differences, though, should not be interpreted as a lack of genetic control. It is not that environmental factors some-how 'outmanoeuvre' genetic instructions, producing differences resisted by the genes. Differences attributable to different environ-ments arise because many genes give conditional, not absolute, instructions. They dictate that if the individual finds itself in one situation it should develop or behave in one way, if it finds itself in another situation it should develop or behave in a different way. For example, people are *genetically* programmed to react to a bad experience – such as being bitten by a dog – by being much more cautious in the future. The difference in attitude towards dogs just described for two clones, therefore, is as attributable to their identical genes as are their many similarities.

When identical twins are raised apart there is much greater scope for different experiences in different environments. The same would be true for clones raised apart. Although they would still share many characteristics, such clones would differ in their appearance and behaviour much more than clones raised together. In the scene, Phoenix-two was born four years later than her elder clone-sister, and hence had a whole suite of different experiences. They were like twins raised apart, and would differ to about the same extent.

Cloning – the Problems

When the first cloning experiment was announced, the popular press was full of futuristic claims about the way that the technique would be used and misused. Macabre predictions were made. The egotistical would produce droves of clones of themselves and take over the world. The wealthy would maintain duplicate sets of all their own organs in case they needed transplants. Fans of famous people would acquire bits of their hero's tissue to create their own clone. This was even extrapolated to suggesting that fans of dead idols would try to bring their heroes back to life. Clones of Elvis Presley were a common vision. Of course, such 'resurrection cloning' can only work as in the scene, using cells that are still alive. Dead cells will not do.

Not surprisingly, such images of the future fuelled public

concern and frightened people into opposing cloning, clouding their eyes to the positive uses of the technology. Religious organizations in particular were quick to comment, responses ranging from reasoned caution to vindictive outrage. Some simply asked for a moratorium on cloning research until ethical questions had been satisfactorily resolved. Others, however, demanded a universal ban on human cloning, suggesting that it should carry a penalty on a par with rape, child abuse and murder. The main concern for all was that cloning would be an assault on human individuality and dignity and that it was wholly unnatural.

We shall discuss the question of human individuality and dignity in Chapter 15. There, we shall also explore the public attitude to 'naturalness'. For the moment, we should concentrate on the more practical problems associated with cloning.

First, the technique is currently notoriously unreliable. Dolly was the only clone produced from an adult cell by the original research team out of hundreds of attempts, and there were miscarriages and abnormal foetuses on the way. However, this is not surprising for such a radical new technique and few can doubt that the success rate will rapidly improve, as it did for IVF and ICSI once a few basic techniques had been learned.

In fact, the incentive for the pioneers to invest much time and money into cloning from adult cells was not great, other than simply to do it. At present, cloning is primarily being developed to produce large numbers of individuals of dairy animals with a desirable trait. High meat or milk yields are two examples. Another aim, once having used genetic engineering (Chapter 8) to produce an animal that secretes a pharmacologically useful substance in its milk, is to be able to produce whole herds of such animals quickly and reliably. All these things can be done faster, cheaper, more easily and more reliably using cells taken from embryos rather than adults. This, then, is where most of the effort is being directed. Cloning from embryonic cells, however, would do little to meet the two main reproductive needs of *people* – alleviating infertility or producing the twin of a child killed by accident or disease as in the scene. Both of these require cloning from 'adult' cells.

Early in 1998, just a year after the original Dolly paper was published, a few scientists even began to question whether Dolly really had been produced from an adult cell. Unfortunately, the original Scottish team had omitted to compare Dolly and her clone-mother via genetic fingerprinting so that Dolly's origins had never fully been authenticated. Also, as Dolly's clone-mother was pregnant at the time, there was a remote possibility that the cell that produced Dolly was not an adult cell from the mother but a circulating embryonic cell from the foetus. The odds against Dolly being anything other than claimed in the original paper were put at about a million to one, but for a while the scientific jury was out trying to reach a consensus. By the end of 1998, however, all scientific tests had been completed and the original research team had been vindicated – Dolly really was the clone she had been claimed to be.

The strongest possible evidence that adult cells can be used in cloning would come from a successful repeat of the Dolly experiment by other research teams in other laboratories. By March 1998 the only positive such development had been a preliminary report in the journal *Nature* from a French team that had a cow pregnant beyond mid-term with the clone of an adult. The clone nucleus was from the skin cell of a two-week-old calf – only a youngster but 'adult' enough to settle the current debate. Then, in July 1998, came reports that mice had been successfully cloned from differentiated cells. Dolly was no longer alone.

Cloning from adult cells will now quickly become a matter of routine for non-human species. The only question that remains is when it will begin for humans.

Cloning – the Future

There can be little doubt that cloning *will* become part of human reproduction – as one of the many and varied options available to future generations. To get it started, though, it may first be necessary to call it something less emotive – artificial twinning, perhaps!

Two situations seem likely to force the break-through, and perhaps the more powerful of the two will be that illustrated in the scene. Who could oppose a mother such as the character portrayed whose beloved child, just three or four years old, is tragically and accidentally killed? The mother's heartfelt and emotional request to have still-live cells taken from her child's body so that they can be cloned and live again as her next child – the dead child's twin – will seem totally reasonable. It will be little different from her asking for the child to be kept alive by any other means that medicine has at its disposal.

The second situation – a little easier to resist, but only just – will involve those infertile couples for whom cloning is the only hope of reproduction; men with no testes, for example, and women with no ovaries. Actually, there will be an alternative for these people which we shall consider in Chapter 11, but it requires cloning technology and in principle acceptance of one technique will require acceptance of the other.

Currently, research into human cloning is prohibited in Australia, Britain, Denmark, Germany, and Spain. Furthermore, nineteen European countries signed a convention in early 1998 in which they agreed to introduce legislation to ban human cloning for *reproductive* purposes. At the same time, Japan's

Council for Science and Technology announced it was setting up a committee to discuss a possible legal ban on human cloning.

The United States has no law at present (1998) to prevent the cloning of humans. There is, though, a long-standing prohibition on the use of federal government funds for research on human embryos. As a result, any work involving IVF, such as cloning, can only be the domain of private laboratories. When Dolly was first cloned, there was nothing to stop such institutions also pursuing the development of human cloning techniques. Rapidly, though, a five-year moratorium on research into human cloning was introduced.

Currently, the United States is at something of a jousting stage on cloning. On the one side, pressure groups such as the National Conference of Catholic Bishops and other religious organizations are pushing for a total ban on human cloning. On the other side, groups such as the Biotechnology Industry Organization and the American Society for Reproductive Medicine are fighting to limit any such ban. Others seek intermediate solutions, such as allowing the cloning of human embryos as long as they are not implanted and brought to birth. This would effectively legalize the cloning of embryos for research, albeit still only in the private sector.

In Britain, there are moves to distinguish between therapeutic cloning and reproductive cloning. Under British law, human embryos less than fourteen days old *can* be used for experiments. If the same law were extended to include cloned human embryos, it would open the way for therapeutic cloning as part of experiments on tissue regeneration. This would greatly benefit organ and tissue transplant technology. For example, a cell could be taken from a person's damaged tissue, cloned and allowed to develop into tissue that could then be transplanted back into the person without the problem of transplant rejection.

The knee-jerk opposition to cloning in the 1990s is very reminiscent of the opposition to artificial insemination in the 1930s,

IVF technology in the 1970s – and in fact is not so very different from the opposition to wet-nursing in the seventeenth century. Whatever actually happens in the first decade or so of the twenty-first century, there can be little doubt that with the passage of time the outcome will be the same. Cloning offers the ultimate solution to the problem of human infertility, and as such, it cannot fail but to have its day. The only question is when. It may take twenty, thirty or maybe even a 100 years of the Third Millennium to happen, but one day cloning will be as accepted a part of the human reproduction scene as bottle-feeding, artificial insemination and IVF are today.

At a rough guess, therapeutic cloning research will be under way by 2000; infertility cloning and resurrection cloning (as portrayed in the scene) by about 2030. What happens then depends on how quickly society moves towards the Reproduction Restaurant referred to in the Introduction.

Part Three

Choosing a Gamete Partner

7

Contraceptive Cafeteria

SCENE 7

One Way for the Poor ...

The woman was angry. It was like talking to a rhinoceros.

'All I'm saying is have a bloody word with her. Anybody would think I was asking you to sail round the fucking world.'

As Maureen spoke, the cigarette in her mouth waved up and down, ash falling on to the hall floor. Her live-in partner was poised at the door, desperate to escape.

'What's the fucking point? She's not my fucking daughter and she'll take no fucking notice of me, anyway. None of you do. She'll make out. Now stop whining and let me go to fucking work.'

Opening the door, he stepped out, then stared back at his partner, looking her straight in the eyes. 'And for Christ's sake will you clean up this fucking pigsty. Do *something* for your fucking keep.'

He slammed the door. In anger and frustration, Maureen threw the wet cloth she had been holding. For a second, it stuck to the door at head-height then slid a few inches before falling on to the leaf-littered and mud-stained mat that pointlessly guarded an equally grubby and threadbare hall carpet. In protest, she left the cloth where it fell.

She hated the winter mornings. It was 6 a.m. – her partner insisted she got up and made him breakfast every morning before he went to work. He had taken out many of the light bulbs so that she couldn't leave them on and waste electricity, and the central heating hadn't worked for years.

Slowly, she climbed the stairs to go back to bed, coughing as she went. It was another hour before her two teenage sons needed to get up to go to school – and in bed she could at least be warm. Taking off her slippers but keeping on her nightdress and dressing-gown, she climbed back into bed. Before settling down, she stubbed out her cigarette in the ashtray on the coffee-stained bedside table.

She didn't sleep, but carried on fretting about her daughter, Tracy, the focus of the argument with her partner. Tracy was the first in any of her or her friends' families to get to university but only hard work and determination had got her there – and only hard work and determination were keeping her there.

Ever since Tracy had been fourteen, Maureen's partner had insisted that she should earn her keep. Whenever there had been casual labour on offer – mainly weekend work, but occasionally also evening or early morning – he had made her take it. Despite this, she'd not only kept up with her studies but had even shone. Then she had wanted to go to university. What was the point, thought Maureen. Why not start earning money the second she could leave school? She had good grades and could have got a good job. Why wait another few years?

But Tracy had been determined and now was paying the price. There could be no respite for her at university. Maureen couldn't help her financially. If Tracy wanted to stay, she had to do it by her own efforts. To her credit, she was succeeding – but at what cost? Work, study, work, study – that's all she seemed to do.

'You'll make yourself ill,' Maureen had told her, the last time they'd spoken.

Maureen sat up in bed and wrapped her dressing-gown tightly across her chest, unable to get warm. Automatically, she reached for the packet of cigarettes by her side and lit up.

It was bad enough Tracy working and studying so hard, Maureen thought. Now there was all this nonsense about her saving up to be BlockBanked. BB'd indeed! Women in their position didn't pay good money to have their tubes blocked. If they looked as good as her daughter, they used their fertility to try to trap the old, rich and lusty into child support. In her day, fertility was the working-class woman's most powerful weapon. If she hadn't conceived Tracy to whom she had, where would the family be now?

Maureen heard a door open across the landing. A few moments later a loud fart echoed round the bathroom. One of her sons was awake, or at least moving. Imagine him saving up to be blocked, she thought incredulously. Mind you, that *was* one of the problems. She could guess what was on Tracy's mind. It was becoming more and more difficult to find a *rich* man who hadn't been BB'd. So, the only men you could trap into child support were those who were too poor to join the BB scheme and so too poor to provide decent support. Look at the man who had just stormed out of the door. She'd conceived to him – twice – and look at the mess she was in now.

A door closed. Her son had gone back to bed. She would have to drag him out, up and to school now he was comfortable – what a difference from her daughter. How much longer, though, she thought, was Tracy going to be her daughter except in name? She'd seen her only once since she'd gone to university, and she was changing already. The crowd she was mixing with was already giving her middle-class hopes and ideals. BB – what a ridiculous notion, Maureen thought, inhaling. Then, cigarette finished, she slipped back under the covers. If only she could get warm.

A hundred miles away, Tracy was also cold. She had taken the cheapest, most basic accommodation she could find – a bed-sit with a kitchen and bathroom she shared with four other people. Huddled in her dressing-gown, she came out of the bathroom and made her way back to her room. It was early and none of her housemates were up. They were all students, like her, but were being supported by their parents.

Back in her bedroom, Tracy took the day's contraceptive pill, took off her dressing-gown, and as quickly as possible tried to cover her nakedness – for warmth, not modesty.

'Christ, Tracy,' came a voice from under her bedcovers. 'Next time, let's sleep at my place. I've got frostbite and my genitals have disappeared completely.'

Tracy laughed as she pulled on her jeans. 'Stop complaining, wimp. Come on, get out of bed and get dressed. I've got to go.' She started work in half an hour, an early-morning shift at a newsagent's.

Reluctantly, Roger dragged himself out of bed under her inquisitive gaze. It was true – he had little sign of genitals. And his breath was actually visible in the cold air of her room.

They met again at lunchtime, in the cafeteria. Tracy had worked for two hours, then had two lectures. Roger had gone back to his warm flat and slept the morning away, recovering from the night's sexual exertions. Not for the first time, he had given his lectures a miss.

'Move in with me,' he said. 'You can't go on living in that igloo of yours.'

Tracy shrugged. 'I can't afford it.'

'You don't need to afford it. You don't need to pay anything.'

'Vagina for sale, you mean,' she said.

'Don't be like that. I really love you, you know I do. I want to sleep with you every night – not just when it fits in. My flat costs me the same whether I'm on my own or not. Save yourself some money. Give up one of your jobs. We could see more of each other.'

Tracy shook her head. 'I'm not going to cut down on working until I've saved enough to be BB'd. It's all right for you. You were scarcely through puberty when you were done. Your sperm are already frozen, waiting for hordes of lucky girls to snap them up when you become famous.'

Roger gave the sort of smile that told her she'd struck home exactly.

'Until then, you can have sex with anybody you like and know you're not going to be trapped,' she continued. 'I want to be in that position. I want to be in control. I'm determined not to make the same mistakes as my mother. It's terrible being poor, you know, sex and babies going together. You grovel around in all these strategies, all these little deceptions, trying to trap this person or avoid that person. You can fancy somebody like hell, but daren't have sex with them because they can't afford to look after any mistakes.'

'Hey, come on,' he said, raising his hands. 'It's not that bad. There's always contraception.'

'What would you know about contraception?' she said. 'Have you ever worn a fucking condom? I bet you haven't. People like you haven't had to wear condoms since they found the cure for AIDS and . . . and . . . Oh hell, those other venereal things. The old ones – what the hell were they called?'

'Syphilis?' he asked hopefully. They were both physiology students, after all. He felt they *should* know, but it was history, not medicine.

'That's it,' Tracy said, 'and the other one.'

Roger shook his head. All he could remember was that he couldn't spell it.

'Anyway, rich people haven't had to wear condoms since their only use became avoiding babies. And what would *you* know about worrying about taking the pill half your life? You're not risking blood clots – or cancer.'

'Messing about with your chemistry is always going to do something else,' he pontificated. 'I don't know how you can so casually stuff yourself with hormones every morning. You're right, you wouldn't catch me taking the male pill, *or* having any of those injections. Look at the side-effects they keep coming up with for *them*. And you're right, contraception is only for people who can't afford BB – but it's better than being trapped into child support.'

He paused, realizing he had just argued against himself. 'OK, so it's rough being poor. But at least *you* don't need to take the pill, not while you're with me, do you? I could afford BB, so I did – and I've never regretted it. Give your body a break. Stop taking the pill.'

'Block, block, bank,' she said, reciting one of the BB scheme's less tasteful adverts in a girlie magazine for men. 'It's as easy as having a. . . .'

Roger smiled. It *had* been a notorious advertising campaign – although there had been an identical campaign aimed at women – except that the 'as easy as' had been replaced by 'easier than'. He still wasn't quite sure why.

Tracy leaned forward to sip her coffee, which was still hot.

'I know,' she said, briefly abandoning her defensiveness. 'I ought to stop taking the pill really. I've been on it five years already. But you know me. I take after my mother. I can't help it – I just *am* promiscuous. I like sex – I just wish it didn't make babies – and so did she. *She* spent her twenties on her back with her legs apart – according to my stepfather, anyway. How the hell she managed to guess who my real father was, I don't know. I think she just pointed at the richest and hoped for the best. That paternity test must have been the high spot of her life. Then she went and ruined everything by getting mixed up with my current stepfather. They've been blowing my child support on booze and fags ever since. If I'm not careful, I'll go the same way. My libido's scared the hell out of me ever since my first grope behind the bike-shed at school. I was only twelve. I just don't seem to be able to say no. Much as I like you, darling,' she said sarcastically, 'I just know that the first time I'm alone and drunk with a half-decent man, I would just do it. I just don't seem to be able to help myself. And you can guarantee that if I come off the pill, the first time I have sex with somebody else, I'll get fucking pregnant – and he'll be poor.'

Roger was hurt. She could have shown some modicum of fidelity. But that was her – a determined, free spirit. It was probably what he loved about her.

'There's always abortion,' he said. 'You don't have to let getting pregnant to somebody who's poor ruin your life.'

'You really don't know anything, do you, with your rich parents, your nice cosy flat, your nice cosy life, and your nice cosy emotions. I don't think I *could* have an abortion. I know women do and I know I seem hard and sensible and single-minded. But if I got pregnant, I don't think I could kill my baby. Anyway, even abortion costs money, and if the guy can't afford child support, he probably can't afford an abortion, either. And I sure as hell can't. Not yet.'

'OK, so don't give up the pill and don't promise you'll be faithful to me . . .'

'I'm sorry,' Tracy interjected, putting her hand on his. 'It's just the way I am.'

'But at least come and live with me. At least come and save yourself some money. Come in from the cold. I want to sleep with you naked – not all trussed up like Eskimos like we were last night. I know my mother has only met you once, but she liked you. If we were live-in partners . . . If she thought we might become gamete partners one day, she might even help you with BB – lend you some money. It's worth a try.'

Tracy shook her head, but then said, 'OK, I *will* come and live with you, as long as you don't behave as if you own me. It could be fun. But I'm not promising anything, and I'm certainly not going to accept any money from anybody – even to be BB'd. I'm going to earn the money myself. I – me – I'm the one who's going to drag myself out of my mother's gutter, nobody else. I've got to prove to myself – and to her – that it can be done. Then I'll feel comfortable about it. Feel I've earned it.'

The pair lived together throughout the rest of their time at university and surprised themselves by staying completely faithful to each other the whole time. Tracy obtained a good degree and stayed on to do a PhD, ironically researching the possible side-effects of long-term blocking in women. But Roger obtained only a modest degree that led to a modest job in a distant town and the pair separated.

Despite all her hard work, Tracy got nowhere near saving enough money to be BB'd while she was an undergraduate. And once she was a postgraduate, she only had time to earn enough money to keep herself. What money she had saved evaporated when her mother developed lung cancer and needed treatment. Without medical insurance, she would have been dead before she got any useful attention on state assistance. Tracy spent a month wrestling with herself over whether to use her hard-earned money to pay for as much of her mother's treatment as she could. In the end, she did – but it was to no avail. Her mother died within a year of the first symptoms, leaving Tracy without savings. Maureen's funeral was the last contact Tracy had with the family of her childhood.

Armed with a PhD, she landed a good job in a biomedical research lab attached to a hospital, and carried on her research into the side-effects of blocking. Not that she ever found any – all experiments came up negative.

Most of the time she lived alone. Her sexual escapades were mainly one-night stands with co-workers, medical students and the occasional doctor. Even though every single one had been BB'd, she obsessively avoided becoming pregnant. She never missed taking her morning pill, and if there was any hint that sickness had stopped one from doing its work, she either avoided sex for the month or insisted on the man using a condom.

Tracy was thirty-three when her life changed. Having nearly saved the money to pay for herself to be BB'd, she fell in love with a consultant gynaecologist ten years older than herself and abandoned her lifelong principle of refusing to accept financial help. Within a month of meeting him, she happily agreed to be his live-in partner, and within a further month agreed equally enthusiastically to be his gamete partner. He had, of course, been BB'd when a teenager, so she would have to have IVF with him, anyway. Eventually she relented and let him top up her savings so that she could be tube-blocked and bank a supply of eggs at the same time.

It was during the pre-banking tests that she was found to have ovarian cancer. Her illness needn't stop her having children of course. They could afford to have gametes manufactured for her, or to have her cloned. But she didn't want a clone. She wanted her baby to be with her chosen gamete partner, and with a natural gamete of her own making – cytoplasm as well as nucleus.

In the end, her ovaries were removed and the few healthy bits that could be found were transplanted into mice. They produced six eggs for her – the nearest she was going to get to making her own eggs. She was too ill to gestate her first baby, so they employed a surrogate, but she insisted on gestating her second herself.

Tracy lived only another ten years, but in her will she set aside money for her two children to be BB'd if they wanted. It was her wish that they should never be distracted as she had been by having to rely on ancient methods of contraception for *their* family planning.

Family Planning: Ancient and Modern

With cloning and surrogate ovaries available, at least to the in-fertile, and the BlockBank (BB) system – to be explained later – in full swing, this scene is probably set around 2050. But apart from her aspirations, Tracy could have been any woman with a similar background from the last quarter of the twentieth century onwards. In particular, she showed two traits that are becoming an increasingly common feature of the modern woman. While striving to accumulate money and status via a demanding career, she used contraception to delay reproducing until she was in her thirties. Then, when she did reproduce, she had two children in rapid succession. *En route*, she involved herself in the age-old search for a gamete partner with whom to reproduce.

These three aspects of human babymaking – age at first reproduction, family size, and finding a gamete partner – are the main subjects for discussion in Part 3 of the book. How many children will women have in the future? How late in life will they delay starting a family? How will they find their gamete partners?

The search for a suitable gamete partner has always been important. After all, success or failure here has always had the power to make or break a person's life. But the search has also always been dangerous, not least because the quest for compliance and commitment often involves sex with a succession of potential partners. Even now, women risk unwanted pregnancies, men risk unwanted child support and both sexes risk sexually transmitted diseases.

Increasingly over the past century or so, people have sought to shield themselves from all of these dangers by the use of contraceptives. There has been an ever-growing demand for efficient but safe and user-friendly forms of contraception, and the wide range of methods that now exist has been one of biotechnology's great twentieth-century achievements. On closer inspection, though, there is still room for improvement, and the BB system featured in the scene could become one of the future's favoured options.

Part 3 contrasts, in its two scenes, the impact that future contraception technology will have on the wealthy and less wealthy echelons of society. At the same time it highlights what is likely to be one of the greatest ironies of future reproduction. This is that IVF – developed initially to help people *make* babies – will eventually become an integral part of the process by which many – probably the most affluent – will *avoid* making babies until they want them. We first encountered this irony in Chapter 4, and now it is time to discuss it at greater length.

Most people think of family planning and contraception as modern inventions. Many of us assume that our bodies are urging us to have baby after baby, and that our only defence is the conscious use of contraceptive technology. It is a shock, then, to discover that without a modern contraceptive in sight our hunter-gatherer ancestors used to give birth to only three or four children in their lifetime and that some – such as the Andamanese – delayed having their first child until the woman was about twenty-eight. Clearly something unexpected is happening, and if we are to guess what might happen in the future we must first understand what has happened recently – and in the past.

A Family Size of Two

The twentieth century has seen a worldwide decline in family size and slowly the mid-century apparition of unlimited population growth is fading. The United Nation's 'medium projection' suggests that by the year 2000 the world's population will be around 6.2 billion (that's 6200 million), increasing to 7.5 billion by 2025 and to 10.4 billion by 2100. After that, it should stabilize at around 11 billion by 2200.

This projection assumes, though, that the current global decline in family size will continue and gradually fall to the 'replacement level' of about two children per woman by 2100. Current trends are encouraging. Thirty years ago, the average woman in the world had six children. Today's average is 3.5. It is lower – and already around 2.0 – in industrial countries but still high in many Third World countries. In sub-Saharan Africa, for example, it stands at six babies per woman; on the Indian subcontinent it is four.

In industrial countries, this twentieth-century decline in family size actually began in the nineteenth – in France even in the eighteenth – century. But in Third World countries, the decline has only just begun. Universally, though, modernizing the environment has caused first a downturn in the death rate and then, a decade or so later, a downturn in the birth rate.

This decline in birth rate is not some global manifestation of failing fertility. It is a biological response, and it is a paradox. When health and survival improve, the best way for a woman to *increase* her reproductive success is to *decrease* the number of children she has. This is because *maximum* success is achieved by producing the *optimum* number of children, not too few and not too many. It's a simple principle, though one that is often misunderstood, and is best explained in terms of the number of grandchildren a woman ends up having, rather than the number of children.

If a woman produces too few children then even if they all survive and reproduce she will not have as many grandchildren as somebody who produced more children in the first place. If she produces too many children, though, the stresses and strains of family life in overcrowded conditions can badly influence the survival, health and fertility – even the attractiveness – of those children. A woman can still end up with few grandchildren. Inevitably, therefore, for any given woman in any given set of circumstances there will be an *optimum* number of children – the number that will give her the maximum number of grandchildren.

Just how many children are best depends on the environment. The greater the infant mortality rate, the more children a woman needs to have just to ensure that some survive. But if the mortality rate is low, she gains from investing as much as possible in each child in order to make each one as healthy, fertile, attractive and successful as possible. Inevitably, this means reducing the number of children so as not to spread the family resources too thinly.

The modern woman in an industrial society, with high confidence that her children will survive to adulthood, rarely feels the urge for a large family – around two children is the average if we exclude those women who are infertile. Our ancient hunter-gatherer ancestors, who enjoyed a similarly low infant mortality, also responded by restricting family size. Around three to four children was the average.

In between these two modern and ancient groups of people, though, our agricultural and pre-industrial ancestors experienced high infant mortality rates and had correspondingly large families. Seven was the average number of children per fertile woman in many Third World countries, for example. Early industrial families – such as in early Victorian England – also used to have large families. Again, seven was about the average but families into double-figures were commonplace. Despite having so many children, though, whole families could be wiped out virtually overnight by any number of childhood diseases, as a visit to any old graveyard will sadly testify.

The conclusion is obvious, and was most recently emphasized by a conference on world population growth in Cairo in 1994. The basic and in many ways the most powerful contraceptive is an increase in the standard of living and a decrease in infant mortality rate. Then with minimal encouragement women naturally respond by reducing the number of children they have. There is good evidence that even in the absence of contraception technology, the birth rate declines in parallel with the infant mortality rate. Modern contraception simply brings the whole process much more under people's conscious control.

Just how low the birth rate will fall in the future remains to be seen, but there is a suspicion that in industrial countries it is about as low as it will go. Even with a minimal infant mortality rate, fertile women may be reluctant to have fewer than two children. Having just one child – especially if she has it late in life – makes a woman's prospects for grandchildren far too vulnerable to accident, creating the spectre of ending life with no descendants. In China, a coercive 'one family, one child' legislation has only succeeded in bringing the average family size down to 1.6, and that includes the infertile zeros. Even with coercion, women are reluctant to have only one child.

Subconsciously, or even consciously, therefore, the future woman's ancient psyche is likely to make her want to have around two children. The next question is, when in her life will she have them?

Starting a Family after Thirty

Towards the end of the twentieth century, more and more women in industrial societies delayed starting their families until after the age of thirty. During the 1980s, the birth rate among women in their late thirties in England and Wales rose by almost half. At the same time, women in their early twenties began avoiding childbirth, the birth rate falling by one-fifth. Nor was this pattern unique to Britain. In the United States, the birth rate rose by one-fifth among women in their early thirties and by nearly one-half among those in their late thirties. And for many of these older women, the pregnancy was their first. Is such delayed reproduction an example of technology defeating nature? Or is it yet another example of ancient biological urges responding to the environment of the late twentieth century by telling women to delay having their first child? The answers seem to be 'No' and 'Yes'.

Biologically, there are both pros and cons to delaying reproduction. Whether delay pays or not depends largely on the balance sheet of the woman's circumstances, and in the late twentieth century, the average balance sheet began to change in a way that biologically favoured delay.

A woman who does delay reproduction gains several benefits. She avoids the stresses and strains of early parenthood that can sometimes have long-term negative effects on her health and fertility. In addition, her greater experience means that her children are less likely to suffer accident or disease. Finally, she – and/or any partner – has more opportunity to gain the wealth and status that will allow her to do the best for her children. But there is also a potential price to pay for delayed reproduction. A woman risks losing her fertility before she can have children – a mistake made by Tracy in the scene, though she was reprieved by technology. In addition, she limits how many children she can have before reaching the menopause. And finally, by the time her children are twenty, she is relatively old – approaching sixty.

The risk to fertility is insidious; it isn't always dramatic and final as in Tracy's case. Older women generally face more problems with conception and birth than younger women. Failure to conceive, complications with pregnancy, and various birth defects in the infant, all climb steadily as a woman grows older. Even so, none of the risks are very great, and they are still declining. For example, the 1990 statistics show that the risk of death to the mother in Britain, over ten times greater for the over-forties than for the early-twenties – translates into only about fifty-four deaths per 100,000 older women compared with five per 100,000 younger women. High blood pressure and diabetes in particular are more common for older women, the former creating a risk for the foetus as well as the mother. None of the risks, though, are as great as just a few decades earlier.

Breech births are more common in older women. So too is the risk of miscarriage. In New York the risk of miscarriage to women under thirty was a little over 10 per cent. It climbed slowly to about 15 percent at thirty-seven, and by forty-five rose sharply to about 45 per cent. Why? The older womb must play some role because older women are more likely than younger to lose foetuses that appear to be perfectly normal. However, the number of foetuses with abnormal chromosomes also rises with maternal age, and the vast majority of such foetuses miscarry.

Most chromosomal abnormalities rise with maternal age because of the increasing proportion of faulty eggs that are released from the ageing ovary. Studies suggest that over the age of thirty-five, about one-third of a woman's eggs are chromosomally abnormal. The risk of Down's syndrome (an extra chromosome 21) and some other chromosomal abnormalities rises with age. At age twenty-five, the risk of Down's is one in 1500; by forty it is one in 100, and at forty-five, one in thirty. In addition to spontaneous abortion, many affected foetuses are deliberately aborted after being detected by amniocentesis and chorionic villus sampling. In 1989, more than two-thirds of Down's syndrome foetuses in women over forty were aborted.

Along with more miscarriages, the foetuses of women over thirty are significantly more likely to die late in pregnancy. Finally, in women over thirty-five, there is a slightly higher risk of a baby with low birth weight or a premature delivery.

At first sight, this list makes depressing reading for a woman who is thinking of delaying reproduction until her mid-thirties, but she should not react too badly. None of the risks are actually very great and for many *modern* women are easily outweighed by the advantages.

Of all the *advantages* of delayed reproduction, the most important are probably the increased wealth, status and experience a woman can offer her children. Children are most likely to survive and grow into healthy, fertile adults if they are born to a competent mother in a favourable environment. Plenty of space and an adequate supply of healthy, nutritious food are paramount. Children then have the lowest risk of contracting diseases and the greatest resistance to those diseases they do contract.

In modern societies, space and nutrition depend on wealth. Even in a country such as Britain at the end of the twentieth century the chances of a child from a poor family dying before reproducing were double those of a child from a rich family. In the historical and evolutionary past, these differences will have been even greater. Theoretical calculations have shown that because of the importance of resources a person is most successful reproductively if he or she spends roughly half the time in direct reproductive activity – seeking mates, spending time with children, etc. – and half the time trying to build up resources. With a few provisos, this is roughly what other primates do – and it is also roughly what humans do. And in the modern environment, the best way to do this is first to acquire resources, *then* reproduce.

So given these traditional costs and benefits of delayed reproduction, what changed in the last part of the twentieth century? Why did ancient urges lead more and more women to delay their first reproduction? And what will happen in the future?

There were three critical factors. First, thanks to huge advances in medical science and infertility treatment, the dangers of delay decreased dramatically in the last part of the twentieth century. Second, optimum family size fell to about two, needing much less time to complete than did the larger families earlier in the century. Finally, more and more women were achieving financial independence; those on their own being increasingly able to support themselves; those in a relationship contributing an increasing share to the household income. More women than ever before, therefore, stood to gain by putting career before family while they built up their wealth and status.

All these factors are set to become stronger, not weaker, in the foreseeable future. The scene is set for women's age at first reproduction to continue to rise as we move into the twenty-first century. Just how late women will decide to leave having their two or so children in the future, though, is a fascinating question. The pressure to delay and delay will undoubtedly grow stronger, and late thirties or even early forties might become a better age to start than the early thirties. At the moment, of course, women are still constrained a little by their declining fertility from thirty-five onwards, particularly by the ever-looming menopause. But all of this is destined to change, as we shall see in Chapter 14. In all probability, women *will* delay having their children until their *forties* in the not too distant future.

Today, the number of women over forty who become pregnant is actually relatively small. As recently as 1964, almost three times as many Englishwomen over forty were having babies as now. The difference is, of course, that whereas in 1964 most of the over-forties who had children already had other children, in the 1990s an increasing number of women were having their *first* child at forty. It is this trend that will grow and grow over the decades to come.

Bottle-feeding and Rapid Reproduction

We have just concluded that the modern woman often needs to have her family *very* quickly after a long delay getting started. In the next section we discuss the pros and cons of modern contraception and conclude that most are unsatisfactory. Yet, paradoxically, the twentieth century saw women in their droves discard one of their most effective *natural* contraceptives in favour of technological alternatives. Once again, though, there is a biological logic linking these apparently disparate contradictions.

Natural selection shaped humans, and many other primates, to be less likely to conceive while lactating. For example, lactating humans may take six or more months to start menstruating after giving birth, and almost always this first cycle while lactating is infertile. Even the next three cycles after that have a less than fifty-fifty chance of being fertile. A study in Chile found that none of the exclusively breast-feeding women had conceived within six months of birth, compared with 72 per cent of non-breast-feeding women.

Why was it so important for natural selection to decree that human females should delay conception while lactating? It didn't *have* to. Many, many mammals regularly conceive while lactating, some even being stimulated to ovulate by the very same hormonal regime that inhibits ovulation in humans. The answer is that women inherited a basic problem from their primate progenitors – it is very difficult to carry more than one child at a time when walking long distances.

Of course the problem was, and still is, most extreme in those human cultures in which women are responsible for collecting and carrying large quantities of food, water, firewood or other materials. Carrying even one child at the same time is difficult and tiring – carrying two would be almost impossible. It was important to avoid having another child until the previous child could not only walk but could also keep up. At least that way a woman only ever needed to carry *one* child while going about her daily routine. The desert living !Kung San of Southwest Africa, for example, have an average birth-interval of about four years. !Kung San women are responsible for day-to-day foraging and often have to carry heavy food as well as their youngest child for long distances. Women who conceive less than four years after their previous child are more likely to damage themselves and their long-term reproductive prospects.

The problem of how to carry two children at once isn't completely unknown even to the modern woman of industrial society – but thanks to cars, prams and pushchairs, having two children close together is far less likely to cause permanent damage than it used to. In fact, having two – or three or four – children *very* close together can sometimes be a very good plan for the modern woman who has delayed having her *first* until her thirties. And as luck or coincidence would have it, her ancient legacies allow her to do so, thanks to modern technology.

By abandoning breast-feeding, the modern woman can rapidly re-install her fertility and be ready to conceive within months if that is what fits in with her family plan. In addition, the adoption of bottle-feeding frees her to spend long periods of time away from her child, which again can be very useful in the modern career environment.

Bottle-feeding, then, offers the woman of the twentieth and twenty-first centuries a number of advantages – enough, it seems, for most to decide that any extra costs in terms of health and finance do not outweigh the benefits. Being forced to use technological contraceptives rather than the free contraception that comes with lactation is evidently a small price to pay, even if those modern methods do suffer all the disadvantages that we can now discuss.

Contraception: Room for Improvement

Life is a roller-coaster of highs and lows, good times and bad, and parenthood is a demanding occupation. The most successful family planning produces children during the good times and avoids having them during the bad. In the past, this meant coinciding reproduction with abundant food, plenty of living space, and a secure society – and avoiding reproduction during periods of starvation, crowded conditions, and times of social turmoil, such as war. The very effective system that natural selection produced for our ancestors was to intersperse periods of *stress-induced infertility* during the hard times with periods of high fertility during the good times.

Until the twentieth century, stress was most women's only contraceptive – as to some extent it was a man's. Stress is still an important factor in fertility, of course, as we discussed in Chapter 3. Nowadays, though, couples naturally prefer to rely less on stress and more on biotechnology – but at a price. As Tracy and Roger pointed out in the scene, virtually every method of contraception that has ever been invented is decidedly 'unfriendly', inefficient or a danger to the user's health.

Barrier Contraception for Females – the Diaphragm

There is nothing new about barrier contraception for women. For centuries, the females of different cultures have been placing leaves or fruit (even crocodile dung!) in their vagina in an attempt to avoid conception.

Such people had no more than the most intuitive of ideas of

how their actions might work, of course. Sperm were not seen down a microscope until about 300 years ago. But many societies seemed to reason that if a baby came out of the vagina it probably got in the same way, so that was the passage to block. Many societies, particularly around the Mediterranean, also seemed to guess that the man's semen was a likely culprit.

Crocodile dung is probably a good spermicide and fruit is probably a reasonable barrier – and in the guise of various creams plus the diaphragm, which fits over the cervix, these are still the two main barrier principles for women today. There are relatively few side-effects, save occasional reactions to spermicide. However, they are fiddly and messy to use, kill spontaneity and although they were once popular in some countries, they are now used by fewer than one in ten women worldwide.

The failure rate of the different barrier methods is relatively low if they are used properly. Of women using just spermicides for a year, for example, about one in twenty would be likely to conceive, whereas if they used a diaphragm as well only about one in fifty should conceive. However, both methods are easy to use incorrectly and then up to one in three users conceive.

Barrier Contraception for Men – the condom

The idea of hindering or killing semen as it leaves the penis is far from new. Over 2000 years ago, Pliny suggested rubbing sticky cedar gum over the penis before intercourse. The sheath has been known since Roman times and was in use in many parts of Europe by 1700. Fallopio designed the first medicated linen sheath in the 1500s but the item took its name from the personal physician to King Charles II, the Earl of Condom, who recommended its use to the king as an aid to prevent the contraction of syphilis. Since then, the only significant improvement has been due to the vulcanization of rubber in the nineteenth century.

By the 1890s, the condom was openly on sale in the UK. However, its use did not become widespread until well into the twentieth century.

The condom has many advantages – it is under the user's control, it takes only seconds to fit, and it helps protect both the man and the woman from sexually transmitted diseases. Maybe, to judge from recent advances, AIDS won't be as big a problem to future generations as was once feared. But new sexually transmitted diseases may well arise to take its place – just as AIDS took over where syphilis and gonorrhoea left off – and despite Tracy's monologue the condom's role in safe sex may never be totally redundant.

With all these advantages it's a pity that the condom is such a sexual turn-off for so many people. Only in Japan is the condom still the most popular contraceptive, largely because the pill is not generally available. But elsewhere, it is much less popular. In the 1980s, only 30 per cent of couples were using condoms in the UK. Worldwide, in the 1990s, the figure was less than 10 per cent.

When used properly, the condom is a fairly effective contraceptive – after a year of its use, only about one in fifty couples should have produced a pregnancy. Again, however, misuse is easy; then the failure rate rises. In a year, as many as one couple in three can produce a pregnancy.

Intra-uterine Devices

The first intra-uterine device (IUD) – basically a small implement inserted into the womb that interferes with implantation – was first developed in the 1960s. In most Western countries it is not a popular option, used by only one in thirty women. In some countries, though, the IUD is the most widely used contraceptive. In China, for example, it is used by more than 40 per cent of women who use modern contraception, making it the country's most common contraceptive method. Most of these IUDs are inserted after the birth of a first child. In Cuba and Vietnam, as in China, women are urged by the government to have IUDs fitted.

Unfortunately, IUDs have a number of side-effects. They can trigger haemorrhaging and have been linked with persistent infertility even after the device has been removed. In addition, they can also be unreliable. Women can unknowingly lose the IUD from their womb and hence embark on what is essentially unprotected intercourse.

Reported failure rates for the IUD in Britain suggest that somewhere between one in fifty and one in twenty-five women would be pregnant after a year of its use. Elsewhere, though, the failure rate can be higher. Of 10 million abortions carried out in China each year in the early 1990s, around 30 per cent were performed on women who became pregnant while using a cheap, locally made IUD.

The Contraceptive Pill

Chemical contraception, as in the pill, isn't a human invention. Female chimpanzees, for example, chew leaves that contain contraceptive chemicals at appropriate times. Nor, for humans, is it a recent invention. Effective drugs were available throughout classical antiquity and medieval times – a whole range of herbal drugs were commonly in use in ancient Greece and Rome. Evidence of chemical contraception stretches back as far as 4000 years to Egyptian papyri. Then, midwives and physicians had many herbs to recommend to women who wanted to prevent conception at the time of intercourse, prevent the implantation of the embryo, induce menstruation or produce abortions. These drugs were sufficiently effective for Juvenal, the Roman satirist, to write almost 2000 years ago that 'we have sure-fire contraceptives'.

Somehow, though, this knowledge was lost to the Western world for a few hundred years. Developed in 1951, the pill has become the most popular contraceptive in most countries. There are various versions on the market, ranging from the mini-pill, which essentially switches off menstruation, to the 'combined oral contraceptive'. This cocktail of oestrogen and progestogen – a synthetic version of natural progesterone – prevents the monthly release of a fertilizable egg.

As might be expected from a system that sledgehammers the female hormone system in such a way, there are a number of side-effects ranging from mood changes, such as loss of libido, to potential health risks. Over the years the pill has been repeatedly, though controversially, linked to potentially fatal blood clots and cancers. For example, young women who use the pill for eight or more years increase the risk that they might develop breast cancer by almost 75 per cent. There also appears to be a link between the pill, smoking and cervical cancer. Tests on women who smoke suggest that the pill increases the damaging effects of cigarette smoking on DNA in their cervical cells.

Alone in the industrial world, Japan banned the pill for contraceptive purposes from the beginning. The initial concern was that its safety could not be guaranteed, followed in the 1980s and 1990s by concern that its use would reduce the sale of condoms and hence promote the spread of AIDS. At the end of the twentieth century, though, there were signs that Japan was about to join the rest of the industrialized world and allow the pill to be used. Women's groups, however, were less keen on the prospect than health professionals – over 70 per cent continued to express concern over possible side-effects.

Used properly, the combined pill has a low failure rate, with less than 1 per cent of women conceiving in each year of use. But it is relatively easy to use incorrectly – such as failing to take other precautions if a pill is forgotten or sickness reduces its efficacy – and then up to one in ten women per year can conceive.

Abortion

Although most contraceptive methods are relatively efficient when used properly, it has to be remembered that they are in the hands of real people with ancient legacies such as less than perfect memories, impulsive sex drives – and a subconscious urge to reproduce. If any testament were needed to the *inefficiency* of modern contraceptive methods in the hands of real people it is the number of abortions that occur every year around the world – although admittedly in some places these are due to a lack of access to reliable contraception. However, every year, about 3 per cent of women between the ages of fifteen and forty-four years have an abortion in the United States – where there is no such excuse. Elsewhere, there are, for example, about 10 million terminations a year in China, 4 million a year in Russia, 300,000 a year in Japan and about 150,000 a year in England and Wales. According to the World Health Organization, half of all pregnancies worldwide are unplanned and a quarter are 'certainly unwanted'.

Laws governing abortion differ in detail from country to country but have overall similarities. In England and Wales, for example, the Abortion Act of 1967 legalized abortion provided two medical practitioners were of the opinion that continuation of the pregnancy involved a risk to the life of the pregnant woman, or to her physical or mental health. In practice, though, this meant that most women could have an abortion on demand. The Act also included the risk of the child being born with serious physical or mental handicap as a legitimate reason for abortion. It did not, however, permit abortion of foetuses capable of being born alive. However, advances in the intensive care of premature infants have meant that the actual age that abortion should be permitted is under continual review. Initial limits of around twenty-eight weeks now seem high to many people.

Abortion has always been an emotive issue. The improved chance of survival of foetuses as premature babies provides a platform for the Pro-Life movement, whose members want to have abortion outlawed, with the possible exception of those pregnancies which present a serious risk to the mother's life. This movement is countered by the Pro-Choice Alliance, whose members call for abortion on demand on the grounds that women should be allowed to exercise full control over their bodies and reproductive performance.

The deliberate termination of pregnancy used to be a dangerous procedure for women, but with legalization and improved techniques, such as vacuum aspiration of the womb contents, the risk has fallen dramatically. In the United States, the risk of women dying during abortion has dropped more than five-fold since abortion was legalized in 1973. Deaths from *illegal* abortions dropped from about 1000 a year in the 1940s to none in 1979. By the late 1980s, deaths from *legal* abortions had fallen to less than one per 100,000 procedures – less than the risk of death from an injection of penicillin and less than one-seventh the risk of maternal death during childbirth. The evidence of psychological damage to women from abortion also appears to be minimal. There is some indication, though, that repeated abortion can lead to infertility.

No matter how safe the termination procedure in modern clinics, many women – like Tracy in the scene – would not contemplate abortion, however unplanned the conception. It is probably true to say that even those that do go through with the process rarely do so with total equanimity. Almost all would say that preventing the pregnancy in the first place would be preferable to termination. Clearly, though, modern contraception techniques are still inadequate – technically, pragmatically and socially. As we begin the twenty-first century there is still a pressing need for a reliable and user-friendly contraceptive system.

The BlockBank (BB) System

Since the contraceptive revolution of the 1960s, there has been little real progress in contraception technology. Variations on old themes are under development – from the female condom and the male pill to various forms of vaccinations – but there is nothing revolutionary. Still none meet the user-friendly, infallible and risk-free criteria that are required of the 'perfect' contraceptive. We might even predict that no barrier system can ever be user-friendly and no system based on altering the body's hormone or immune system can ever be free of health risks.

Part of the problem for contraception technology has always been the need for users to retain their fertility so that they can reproduce later if they wish. Now, though, modern reproductive technology could make this requirement redundant, thus paving the way for a completely new approach to contraception and family planning.

In 1994 in the journal *Nature*, two US scientists – one being Carl Djerrasi, the Californian chemist who made the first oral contraceptive – suggested that men should consider a new form of birth control. They should bank their sperm when they are young – and then have a vasectomy. When they want to have children, they should simply draw on their account and achieve conception through artificial insemination.

As hinted in Scene 4 and now just elaborated in Scene 7,

advances in reproduction technology since 1994 allow us to elaborate this scheme considerably. In the account that follows, I shall use the word 'block' and its derivatives to mean any blocking of the vas deferens or oviduct, whatever the mechanism – cutting, tying, etc. And I shall use the word 'bank' and its derivatives to mean any cryopreservation (freeze-storage) of gametes. Hence, the BlockBank system, or BB for short.

The BB scheme of the future will be very different from that suggested in 1994. First, women as well as men could use the scheme because, since 1994, egg storage technology has been developed (Chapter 3). Second, women who use the scheme will need to reproduce by IVF rather than by intercourse or artificial insemination. As portrayed in the scene, therefore, men and women could be blocked and bank their gametes early in life. By so doing, they would guarantee avoiding unwanted pregnancies, abortions and child tax. At the same time they would give themselves total freedom to have a family at any time they wished. To have a baby, all a person would need to do is arrange via IVF to have their gametes united with those of the person of their choice. The age-old search for a gamete partner would take on a radical new meaning.

To this extent, the system would be the perfect form of family planning – no unwanted pregnancies, and babies to order. In addition, there should be few of the side-effects associated with modern contraceptives – in fact, to judge from people who have already been blocked, there may even be *no* side-effects.

As far as a man is concerned, blocking the vasa does no more than prevent sperm from being ejaculated. It has no effect on his hormonal system and so has no influence on his 'maleness' or moods. Sexual response is as intensive as before, seminal fluid continues to be produced by his prostate gland and seminal vesicles and he continues to ejaculate normally – it's just that his semen contains no sperm.

Similarly for a woman, blocking of the oviducts has no

influence on her hormonal femaleness. A blocked woman still has menstrual cycles. She still ovulates, still has periods, and her moods continue to follow their normal, complex cycles. She can also gestate and give birth to a baby – and breast-feed. All that blocking does is prevent the egg from being fertilized and/or reaching the womb.

Unlike other modern forms of contraception, blocked tubes are 'natural' in at least one sense – they do occur naturally! Around 3 per cent of men and women acquire blocked tubes via urinogenital infections as we described in Chapter 3. And until such people discover that they are infertile, they have no indication from their libido, sexual performance or general behaviour that their tubes *are* blocked.

Relatively few aspects of tube blocking have caused concern. One worry is the possible link between vasectomy or oviduct ligation and depression. However, even if this link is real, it probably owes more to the psychology of self-enforced infertility than to any direct chemical or physical influence of the blocked tubes themselves. Otherwise, people with naturally blocked tubes would show a similar effect.

Another worry is the way that people with blocked tubes continue to produce gametes. Of course, such continued production is an important feature of the 'normality' of such people. Nevertheless, as neither sperm nor eggs ever get shed to the outside, there is some concern over what happens to them. Eggs die in the oviduct – or perhaps in the woman's body cavity – and need a little 'clearing up'. Compared with men, though, the amount of material involved is minuscule. Most sperm end up being ingested by white blood cells, which the man's body has to increase in number to cope with the extra demand. Sperm can also leak into the body cavity, again requiring the attention of white blood cells. Although blocking has no influence on hormonal systems, therefore, it does have some impact on immune systems, particularly – and maybe only – men's. The lack of sperm passing through the prostate gland might also matter and the mooted possibility of a link between vasectomy and prostate cancer has still not been ruled out.

Despite such concerns, there is no clear evidence to date that blocking tubes, whether naturally or surgically, has any *medical* effect other than infertility. There is, though, an occasional but undeniable *psychological* problem. Although many surgical procedures should be reversible, there is still an aura of sterilization about the process. To some people, blocking seems like the end of fertile life – hence maybe an occasional link with depression. Currently, also, although some men take pride in – or even exploit – their vasectomized state through the wearing of ties and badges, many more believe that a vasectomy will somehow or other interfere with their 'manhood'. Some even see it misguidedly as castration – removal of the testes – which would indeed hugely affect their maleness, because it would remove the major source of male hormones.

BlockBanking, though, should engender an image quite different from sterilization. In fact, psychologically, it could gain the aura of the beginning of a person's fertile – and particularly sexual – life; the equivalent of the sexual 'rite of passage' found in so many cultures. As such, it should appeal to many, many more people than do its current equivalents of vasectomy and oviduct ligation. Nevertheless, there are still problems that we have not considered, and some critics have claimed that these are so great that the chance of any such scheme becoming a popular approach to birth control is zero. So, what are the major problems – and are they real or simply a reflex distrust of anything new?

First, as with any banking system, men and women might not trust the gamete banks with their sperm and eggs. What if people's gametes get mixed up or cannot be traced when needed? What if a person falls behind in the payment of his or her gamete storage premiums or if the Storage Company goes out of business? In theory, of course, thanks to DNA fingerprinting, bar code labelling and computer technology there should be little danger of gametes being lost or mixed up – but past experiences with child support agencies, pension schemes and even banks do not inspire total confidence. Payment protection policies would undoubtedly be needed, as would government underwriting of any private storage companies.

None of the practical problems with *banking* are insuperable. It is the *blocking* and babymaking part of the BB scheme that is likely to be the biggest hurdle to its adoption. Currently, the system undoubtedly lacks widespread appeal – particularly for those who dislike all forms of medical invasion. Tube blocking, both male and female, requires surgery, albeit minor. In addition, women would need to go through induced ovulation and egg harvesting when initially banking eggs. Finally, the fertilized embryo needs to be inserted directly into the woman's womb – another uncomfortable procedure. All these processes are unpleasant and such discomfort is a definite minus on the balance sheet of user-friendliness.

Of course, people don't have to bank their gametes at the same time as they are blocked. In fact, they could avoid banking their gametes altogether if they wished, yet still enjoy the contraceptive protection of being blocked. Gamete collection could be delayed until a person wished to reproduce. Blocked men could have their sperm collected by TESE (Chapter 4) and blocked women could have their eggs harvested in the normal way. For women, this would mean they could avoid the discomfort of harvesting until they had the incentive of wanting a baby. For men, though, TESE involves discomfort that they could have avoided had they banked sperm before being blocked.

Against any perceived advantages, women could suffer a disadvantage from not providing eggs until they want a baby. Eggs collected from younger women are more fertile and less likely to have chromosome abnormalities than eggs harvested from older women. To a lesser extent, the same may also be true of men and their sperm. In addition, delay means that both sexes risk being robbed by disease or accident of their ability to produce gametes – just as Tracy was nearly robbed in the scene. All in all, everybody would be well advised to bank their gametes at as young an age as possible, preferably at the same time as being blocked.

Will the men and women of the future be prepared to put themselves through such discomfort in the name of contraception and family planning? Past behaviour suggests that at least some might. In the pursuit of health, women have been prepared to suffer the discomfort and indignity of cervical smear tests for years. In the pursuit of family planning, they have been prepared to tolerate the discomfort of being fitted with IUDs – and later having them removed. They have even been prepared to risk their lives with back-street abortionists. And in the pursuit of reproduction, *infertile* men and women have been prepared to put up with the discomfort of TESE and IVF. Finally, in the pursuit of contraception, both men and women have already been prepared to have their tubes blocked – in surprising numbers. In Britain 15 per cent of women of childbearing age and 16 per cent of their partners are sterilized, with most men having their vasectomies in their thirties. In Asia, half of all couples who need contraception choose sterilization of one partner; in India alone this figure rises to three-quarters. Of course, in Asia many long-term contraceptive acts date back to the dark days of coercive population control: of village committees press-ganging women to have IUDs fitted and of bribes of transistor radios for men willing to be sterilized.

It seems highly likely, then, that if the BB scheme were on offer some people *would* opt to use it in the hope of foolproof and risk-free control over their reproductive lives. They would tolerate three or four moments of discomfort in their lifetime for the freedom to reproduce when and with whom they wanted with no risk of accident or misfortune. It seems unlikely, though, that the BB scheme would take off to the extent portrayed in the various scenes in this book without some reduction in the currently associated discomfort. Most *fertile* people would still probably prefer to run the gauntlet of unwanted pregnancies in return for a continuing ability to make babies via intercourse.

For men, though, the situation could soon change. Blocking will become user-friendly; research into non-surgical vasectomy is already in progress. In the future, for example, blocking might be achieved via laser technology. Alternatively, it might be possible to harness the natural propensity of some disease organisms to block sperm tubes. These organisms could be rendered benign by genetic engineering, then employed to do the same job that they have been doing for millennia. Blocking could soon become a simple process, requiring no more than a vaccination. And for men, as far as banking is concerned, it really is as easy as masturbation – as the publicity campaign said in the scene – for that is all it entails. The BB system could rapidly become a very attractive form of contraception for men, freeing them from the spectre of unwanted and unaffordable child tax.

Such men's partners – if they haven't been blocked themselves – could conceive via artificial insemination of his banked sperm, perhaps performed by the man himself. If they have been blocked, these partners will have to conceive via IVF. Blocked men who need TESE of some description to retrieve fresh sperm will always need minor invasive surgery, and their partners would only be able to conceive via IVF.

For women, blocking can become as user-friendly as for men, because the same non-invasive techniques can be used. Placement of embryos in the womb, though, will always be mildly invasive – if only for a surrogate. On current technology, so too will the harvesting of eggs. There is a potential future development, though, that would do away with even this discomfort. Before long, it will be possible to manufacture gametes from any cell in the body (Chapter 11). This would mean that instead of having *eggs* harvested, a woman would need only to donate a few cells from surface tissue, then have them cultured and stored (Chapter 6). The discomfort of having such cells collected would be no greater than a minor biopsy. Moreover, it would need to happen only once. When technology reaches this stage, the BB system should appeal to a large section of even the female population.

In the scene, Tracy wanted her children to join the BB scheme as soon as they were old enough to agree to the process. If the scheme were available in the future, would any parent really behave like Tracy and have their children blocked – sterilized – at puberty? Perhaps the only people prepared to do such a thing would be those few malevolent parents who, in past times, were prepared to have their son castrated so as to preserve a beautiful soprano voice? Or maybe it would also be those much more numerous parents who, through religious or cultural belief, are prepared to circumcise their son or daughter at birth – or even at puberty? Or maybe it would be the multitudes that nowadays urge their pubescent daughters to start taking the contraceptive pill, despite the long-term risk of thrombosis, breast cancer and – if their daughter ever smokes – cervical cancer.

The chances are that if the BB system ever gains public confidence, many a parent will urge their children to join the scheme as soon after puberty as possible.

The Contraceptive Cafeteria

The World Population Conference of 1994, held in Cairo, encouraged the development of what it called the 'contraceptive cafeteria'. It foresaw the next leap forward in controlling world population hinging on a much greater choice of contraceptives, coupled with more emphasis on what consumers want. The inspiration for the cafeteria approach was its spectacular success in Bangladesh – a poor, rural, Muslim country once viewed as unpromising territory for progress in family planning. But use of contraceptives shot from 7 to 40 per cent within a decade. The key was the provision of a wide range of contraceptives, allowing people to choose what type they wanted, instead of having a doctor prescribe them as if they were medicines.

There will undoubtedly be a wide range of contraceptives on offer in the twenty-first century, including the BB system. People will be able to select the contraceptive that suits their situation, character – and finances. The BB system will undoubtedly be one of the more expensive, although, as a matter of perspective, it will probably cost less than the average 'white' wedding which in Britain currently averages about £10,000. But as the appeal of the BB system will be mainly to career women seeking freedom from pregnancy until their thirties and to wealthy men vulnerable to the predatory seekers of child support, cost will hardly be a barrier. The less wealthy echelons of society, though, will remain limited to a range of contraceptives similar to those available now, with abortion as the last resort. Like the fast food on many a cafeteria menu, many of these contraceptives, as now, will carry health risks, but at least they will be affordable.

The result will be that there will be many Tracys in the world of the future – people who for most of their lives can only afford the contraceptive equivalent of burgers from the contraceptive cafeteria. Nevertheless, they can still aspire to the gourmet meals on offer from the Reproduction Restaurant next door, as described in the next chapter.

8

Reproduction Restaurant

SCENE 8

. . . and Another for the Rich

Meal over, Nathanial and his mother sat down at a computer terminal in a Reproduction Restaurant, each with a coffee by their side. It had been a disappointing season – a quarterfinal place was the best Nathanial had managed in any of the Grand Slams – and he was tired, physically and mentally. He was more than ready for the Christmas and New Year break. For just two or three weeks, he wanted to forget about training, laze by tropical swimming pools, and mildly abuse his body with rich food, alcohol and a succession of ardent fans. The last thing he felt like doing was to give serious consideration to becoming a father.

'Double-click on "Eggs",' his mother said, 'and type "Tennis Players" into the keyword box.'

Of course, his mother had arranged everything. She always arranged everything, managing his life just as obsessively – and effectively, he had to admit – as she had managed his career. It was she who had put a tennis racket in his hand when he was only three, paid for the best coaches, and steered him through to where he was now. Not once had she doubted that, with his pedigree, he would go right to the top.

'Right, now where it says "Arrange by", double-click on "World Ranking",' she continued.

'My God,' he exclaimed, as the list unfolded before him, with tiny thumbnail portraits of the women by the side of each name. 'They're nearly all here.'

'Well, of course. How else is a busy woman like a professional tennis player supposed to have her children? That's why you're here.'

'Mother, I'm only twenty-four. I've got plenty of time to have children.'

'But why wait? Look, we've been through all this, let's not waste time going over old ground. You're not going to see any more of your child than you want. You can afford it. And, believe me, you'll enjoy it . . . Anyway, it's not *your* age that matters. It's mine. Don't be selfish. Come on. Concentrate. Let's look at the list.'

He sat back and stretched. 'Look, Mother. I know you're really keen on your first grandchild being a tennis player . . .'

'Grandson,' she said hopefully, though this too was old ground.

'No, mother. I'm not budging over this. I want the eldest to be a girl. If I do end up having to look after a family by myself, a girl will be more useful. Anyway, I can see myself living alone with a daughter more than I can with a son. I'd like a son later – you can have your grandson, don't worry – but I want a girl first.'

He paused. 'OK, look, let's compromise. Let's have a girl tennis player.'

His mother sighed. 'Well, as long as I can have a grandson before too long,' she said grudgingly. 'I'm fifty-four already. I don't want to be too decrepit when he starts. I want him to see that even Grandma can play a bit. Anyway, we're wasting time. We're paying by the minute, here, you know.'

'We can afford it,' he said. 'You just said so. Otherwise we wouldn't even be here.'

He began to scan down the list of women. 'You know, this is really weird. I know most of these so well. I've even slept with some of them. If you hadn't talked me into being blocked at sixteen, you might even have had your grandson already.'

'I very much doubt it,' said his mother. 'No serious tennis player would take a chance on that. I bet none of these have banked without blocking. In fact, I bet they've all been blocked even if they haven't banked.' She paused. 'Did I get that right?'

'Yes, I suppose so. In fact, I know she's been blocked,' he said, pointing at the current world number one. 'She told me.'

He scrolled down the screen. 'Oh my God,' he said suddenly, as he came across somebody he knew particularly intimately. 'What's she going to think if I ask for one of *her* eggs.'

'She'd probably be very flattered,' said his mother.

'It would be so embarrassing if she said no.'

'Anyway,' his mother continued. 'I don't think she's a very good choice. She's too low down the rankings.'

'But she's on the way up, just like me. I *know* she'll go all the way.' He smiled at his double entendre, remembering a very steamy night in Australasia only ten days earlier, but his mother ignored him.

'I think we should stick to people who have already proved themselves,' she said. 'That's what I did. Your father had been number one for two years when I bought his sperm. They were expensive, but I've never regretted it – look at you.'

He scrolled up the list again. If he was honest, he'd never regretted his mother's actions, either. He'd never met his father – come to that, neither had his mother – but he felt he knew him. The house where he had lived alone with his mother for most of his childhood was full of photographs of his father and cut-out newspaper headlines announcing his most famous victories. And along with the sperm his mother had purchased came a CD with a lengthy biography and nearly six hours of movies of the man from childhood to adulthood. It was strange to think that hundreds, maybe even thousands, of children worldwide had grown up with that same CD as a father. Two of his main rivals on the circuit had turned out to be his half-brothers.

'Hadn't we better check parentage?' he said. Somehow he felt that sleeping with your half-sister, as long as you'd both been blocked, was one thing – but to choose them as a gamete partner might be something else.

'No need,' said his mother. 'When you typed in your ID code the computer automatically excluded everybody with the same parents or grandparents as yourself. The GMB can be prosecuted for promoting incest, you know, even if it is an accident.'

Reassured, he concentrated on the ten women on the screen who were ranked in the top twenty.

'What's wrong with the number one?' asked his mother.

'She's too big, too ugly and has a bad attitude. Anyway, she's too expensive. Look at what she's asking, and that's per egg, not per baby.'

'Never mind the price – and what does it matter whether she's ugly or not? You're not going to have to have sex with her. It's her genes you're after.'

'Well, it matters to me. I want to have an attractive gamete partner. Don't ask me why, I just do. Besides, I don't want an ugly daughter. If I'm going to have to share my house with her, I at least want her to be pretty. What about number eight?' he said, pointing at the only person in the top ten he had slept with. The thought of having a baby with somebody he'd actually had sex with quite appealed to him.

He double-clicked on her portrait and the screen filled with a portfolio of photographs and brief biographical details. Typically of her, there were even three nude poses – from front, back and side – which left nothing to the imagination. If his mother hadn't been there, he would have double-clicked on everybody by now, just to see if any of the rest had included such shots.

'You see,' he said, 'she's got everything.' He knew that anyway, from personal experience.

'Mmm, not bad,' said mother. 'A bit on the skinny side – but I do like her style on the court. It's her temperament that bothers me, though. How often have we seen her crack under pressure?'

She read some of the text. 'Anyway, it says here that her eggs need screening for the cystic fibrosis gene, so she's more expensive *and* they've run out of her eggs, anyway, and won't be getting any more until June. We can't wait that long. I'll be surprised if you win a Grand Slam this year – but next year . . .'

'What's my winning a Grand Slam got to do with becoming a father?' he said, reluctantly closing number eight's portfolio. Without waiting for an answer, he continued, 'How about number five, then?' He brought up her portfolio on the screen and liked what he saw. No nude shots though. 'I've scarcely met her and never spoken to her – in fact, her English isn't very good – but everybody says she's really nice and lots of people are tipping her to be a future number one.'

'What do you mean, what's winning a Grand Slam got to do with becoming a father?' She was irritated that he seemed to have forgotten virtually every conversation they'd ever had on the matter. 'You know full well what it's got to do with becoming a father. I *do* want to be there when you win your first major, you know. It would be just my luck if you did it while I'm giving birth. But if we time it right, we can make sure I've had the baby before that Wimbledon – preferably even before that French, though I don't think your first title will be on clay. So we can't wait another six months to get the process started.'

'I've told you before. The answer's simple. Don't be the surrogate. Let's pay somebody else to do it.'

His mother shook her head. 'No. *I'm* going to gestate it. I'm not going to risk some hormone-heavy stranger suddenly deciding she's going to keep my granddaughter after I've had – sorry, after *we've* had – nine months of anticipation. They're still allowed to do that, you know, even now.'

'Yes, all right,' he said, resignedly. 'OK, so what about number five then?'

Future Nightmares or Ancient Realities?

Nathanial has just sat in front of a computer screen in one of the Reproduction Restaurants described in the Introduction and calmly tried to choose a gamete partner. His aim was a tailor-made daughter for himself and a tailor-made granddaughter for his mother. The process was unemotional, calculated and blatantly eugenic. More than that, Nathanial – just like his mother, twenty-five years earlier – was able to enjoy this option simply because he could afford it. In the previous scene, Tracy had been prepared to work hard throughout her adult life to try to gain a similar privilege. Is this the way of the future – say around 2075?

One of people's biggest fears over the new reproductive technologies is the way they will change traditional ways of doing things. This chapter focuses on two of those fears – the divorcing of sex from reproduction and the gradual infiltration of eugenics into day-to-day decisions. As always, though, what is one person's nightmare is another person's fantasy – and even future nightmares can sometimes emerge on analysis to be little different from ancient reality.

Divorcing Sex from Reproduction

Nathanial's mental ability to divorce sex from reproduction is nothing new. It has even been said that such ability is a unique human trait – that we are the only species for which sex is recreational as well as reproductive. The human psyche has *always* been able to divorce sex from reproduction, the two seeming so unrelated that many of our ancestors had more trouble making a connection between the two than separating them.

The main difficulty in spotting a link between intercourse and pregnancy arises because people have intercourse far more often than they become pregnant. With an average of about one conception for every 500 inseminations, a relationship between the two can seem absurd – and some ancestors never managed the connection at all. Crocodile dung in the vagina was as likely to be aimed at preventing spirits from going any further as it was to prevent semen. Indigenous Australians, Brazilians and many Africans thought that babies entered the mother from the environment, such as while swimming in water. Other Africans thought that babies were formed entirely from menstrual blood accumulating in the female. Sex didn't enter the equation.

Various other societies realized a man and sex were involved in making babies somehow, but thought that the man simply primed the female's body for conception, then played no further part. The ancient peoples inhabiting the Mediterranean basin – Egyptians, Greeks, Romans – showed more insight than most (Chapter 7), but even they remained bemused by the actual mechanics. A common view was that whole babies originated and were 'incubated' in the man, perhaps in his brain, then passed to his penis before being 'seeded' into the female in the seminal fluid. Aristotle was one of the major supporters of this seeding hypothesis – and it was still going strong by 1700. When sperm were first seen down a microscope about 300 years ago, scientists really thought they could see tiny whole humans in human sperm, donkeys in donkey sperm, and so on. The entities were hence named spermatozoa, which means 'seed animals'.

The first European to record how male *and* female might contribute to a baby via sex seems to have been Hippocrates. He suggested that the menstrual blood, once it stopped flowing out of the vagina, accumulated in the mother to form the baby's flesh, the seminal fluid then forming the brains and bones. As a result, people thought that the end of menstruation was the most fertile time. It was not until the late 1800s when sperm were first seen fertilizing eggs (in starfish and sea urchins) that the process began to be understood. Even then, it was the 1920s before it was realized that mid-cycle was the most fertile time.

Even societies – such as our own – that recognize the link between sex and reproduction nevertheless acknowledge that sexual urges are very separate from, and much more frequent than, the urge to reproduce. Hence the idea of sex as a recreational activity.

To the evolutionary biologist, there are good reasons why natural selection has shaped humans to seek sex for its own sake even though there is rarely a chance of reproduction. The ancient urges that lead a woman to seek or allow sex every few days were shaped for sexual crypsis (Chapter 2). And because a man is unable to guess when his partner is fertile, his only subconscious option is to try to maintain a continuous trickle of fertile sperm through his partner's oviducts. Sex every few days does just that. Hence natural selection has shaped both men and women to feel like the gratification of sex every few days – even though most of the time conception is not on the agenda.

Humans are not the only species to seek sex far more often than is needed to reproduce. Lions, for example, have sex 3000 times for the production of each lion, and some birds have similar excesses. Our two nearest relatives, chimpanzees and bonobos – the latter in particular – are forever having sex. For them, sex is almost a currency, with very few offspring to the dollar. It is nonsense to claim that humans are the *only* species for which sex has a recreational as well as a reproductive role.

Clearly, then, it is not modern technology that has separated sex from reproduction in the human psyche but natural selection.

Such a dissociation of cause from effect is one of our most ancient legacies. All that technology has found the potential to do in the future is reduce the link between sex and reproduction from 0.2 per cent (1/500) to 0.0 per cent – something natural selection could never quite have managed.

In the scene, Nathanial had as active a sex life as any man might in his position. Contemporary technology had not robbed him of any of his urges or gratifications – nor had it robbed any of his female partners of theirs. The only difference between now and then is that, when it comes to fathering a child, his sperm were destined to meet an egg by some means other than intercourse.

Eugenics – or Mate Choice?

On reading the discussion between Nathanial and his mother many people will have felt distaste – and maybe even fear – for a future that could spawn such a situation. A major element in people's reaction, though, will not be the scenario, but a word. Who could read that story without thinking 'eugenics' and reacting accordingly?

'Eugenics' is one of the most frightful, explosive words in the whole bioethical dictionary. Yet the process is an ancient legacy, part of normal everyday life. It never went away and it cannot be excluded from the future. The only question is – what form will it take?

The word itself – meaning 'well born' or 'of good heredity' – was coined in 1883 by Francis Galton, Charles Darwin's cousin. Galton proposed to move evolution from theory to practice and designed a surprisingly mindless programme for breeding a superior race of human beings. Eugenics became wildly popular. Right-wing social Darwinists and left-wing socialists promoted it with equal enthusiasm.

On arrival of the concept in America, generous donors funded research centres such as the Race Betterment Foundation in Michigan. Eugenics went berserk in the United States long before being adopted by the Nazis in Germany. It fed on the American enthusiasm for progress and grew fat on the national fear that further immigration would contaminate the established gene pool. The result was a burst of programmes for compulsory sterilization of criminals and mental patients, restrictions on immigration, and laws prohibiting interracial marriage. Although the programmes have long since ended and research institutes closed down, some of the offensive programmes from that era are only now coming to light.

On the crest of the eugenics movement, Elizabeth Nietzsche – sister of Friedrich – established a colony in Paraguay that she called Nueva Germania. She populated her paradise with splendid men and women, chosen for the 'German purity of their blood', and encouraged their selective breeding towards a race of supermen. The results – still visible in that area – are blond and blue-eyed Paraguayans, many of them now poor, inbred, and diseased.

Unpleasant memories such as these – but particularly the horrific perpetrations in the name of eugenics during the Nazi era in Germany – have guaranteed a universal reflex to any hint of genetic discrimination. This reaction is a pity. In fact, it is more than a pity. It has become a huge obstacle to sensible discussion because, like it or not, reproduction has *always* been a process of eugenics.

In the scene, Nathanial and his mother were doing no more than they would have done now or in our evolutionary past. They were sitting discussing prospective 'mates' for the son and potential 'mothers' for the woman's grandchild. Is this eugenics in any more fearful a form than humans have been indulging in for millennia in the guise of mate choice or arranged marriages?

Humans choose their mates – or their children's mates – and show distinct likes and dislikes in the process. The underlying principle is simplistic and straightforward. We are pre-programmed to be attracted to features in a potential mate that indicate health, fertility and just generally 'good genes'. The evolutionary logic is that those 'good genes' will then be passed on to any children we might have.

The showing of preferences during mate choice is not just a human trait. Every animal studied shows clear preferences when choosing a mate, as Darwin was one of the first to stress in his 1871 book *The Descent of Man and Sexual Selection in Relation to Sex*. Males are less choosy than females (Chapter 2) but even they find some females more attractive than others. Both sexes appraise and respond to the quality of potential mates' genes. Throughout the natural world, mate choice *is and always has been* a eugenic process.

So when Nathanial weighed up the qualities of the women available to him, he was doing no more than men and other animals have done – openly, mentally or subconsciously – for millennia. True, he was sitting in front of a computer, looking at pictures on a screen rather than viewing his potential mates in the flesh or in his mind. And, because he was interested only in reproduction and had no need to fret about compatibility, he was more concerned with their genes than with how good a companion or mother they would make. These, though, are mere details – his approach was ancient.

Nathanial's mother was as interested in her son's choice of gamete partner as he was. She, though, was less concerned with looks than tennis ability. Such parental involvement is also antique. Thanks to humankind's long life-span, overlap of generations and past tendency to live in extended families, people have always taken an intense interest in their children's choice of partner(s). After all, the number and quality of a person's grand-children is directly influenced by the quality of the gamete partner(s) acquired by their offspring.

The discussion between Nathanial and his mother was an ancient scenario, not a future impropriety. Eugenic it may have been, but it was nothing new.

Eugenics, the Human Genome Project and Gene Therapy

Most of the discussion between Nathanial and his mother con-cerned female qualities, such as looks and abilities, which would be obvious to any observer – past, present or future. The abilities are basically genetic but are clearly overt. Just once in their discussion, though, the couple touched on a quality that was not overt, and until recently could only have been guessed at. One of the potential partners – number eight – carried the gene for cystic fibrosis and seemed obliged to announce the fact on the Internet. Nathanial and his mother rejected her as a choice – not as it happens because she carried the gene but because she had no eggs in stock. Some people, though, *might* have rejected her as a gamete partner because of her gene.

In some cases a single abnormal gene from one parent is sufficient to cause disease. Such genes are said to be *dominant*. More commonly, a child must inherit a defective gene from both parents before the disorder shows itself. Such genes are said to be *recessive*. Healthy people who have just one copy of a defective recessive gene are called *carriers*. They do not suffer from the disease themselves but could pass it on to their children. We all carry a few potentially harmful recessive genes but are unaware of them – so too are our gamete partners. Future technology, though, could change all of that. It might allow certain knowledge of a person's genes to become part of mate choice, as we have just seen.

Choices based on such direct knowledge are in principle no more or less eugenic than choices based on the indirect knowledge that comes from physical attraction. But today, as genetic information becomes more accessible, more and more people make eugenic decisions that are based on certain knowledge. It is after all eugenic to abort a foetus with chromosome defects so as to try again. It is also eugenic for a sperm bank to screen prospective donors to find what traits they carry – and reject some on the basis of what they discover. Yet both these procedures are commonplace. As many people have pointed out, there may well be more real eugenics going on today than when it was popular in the United States or Nazi Germany.

As an example, the *New York Times* ran a story in 1993 about a community of Orthodox Jews who have a programme of genetic testing for young people. The goal of the programme is simple and laudable: to eliminate common inherited diseases such as Tay-Sachs. Among Ashkenazi Jews, one person in twenty-five is a carrier of the Tay-Sachs gene. The carriers do not suffer the disease itself – it is recessive – but when two carriers marry, there is a one-in-four possibility in each pregnancy that the child will inherit a copy of the gene from both its mother and its father. It would therefore suffer from the disease itself – and suffer is the word. Tay-Sachs is an incurable, fatal disease in which the child eventually becomes blind and paralysed.

Every year, representatives of the Committee for Prevention of Jewish Genetic Diseases go to the Orthodox high schools and offer the students a blood test. Those tested are given an identification number that is registered at the programme's central office.

Later, when a boy and girl are being considered as likely prospects to be united in marriage, one of the first steps is to check their identification numbers. The records then show either that the match is compatible or that both carry a recessive gene and so would be likely to produce children with Tay-Sachs.

The members of the community were apparently quite satisfied with the programme, which the religious leaders had named Dor Yeshorim, Hebrew for 'the generation of the righteous'. The results are impressive: 'Today,' one report states, 'with Dor Yeshorim's continual testing, new cases of Tay-Sachs have been virtually eliminated from our community.' The programme was being expanded to test for several other diseases. There is no disguising the irony that the ancient Jewish tradition of match-making is now overtly eugenic.

Selective matchmaking – whether by community elders, parents or by a young couple themselves – has the potential to avoid the birth of people who suffer from genetic disease. Matchmaking, though, does not reduce the frequency of the gene in the gene pool. In fact, it helps to maintain that frequency. People still carry the gene but each person has only the single dose and so does not suffer the disease.

Very soon, though, biotechnology will do what matchmaking cannot. It will provide the means to rid the human population of genetic diseases – at least those caused by a single gene. It will do so first, because the Human Genome Project will pinpoint the location of each such gene among the mass of human DNA, and second, because gene therapy will replace each defective gene with one that does not cause disease. People will then have a whole new set of eugenic decisions to make.

The Human Genome Project

The Human Genome Project is perhaps the boldest biological endeavour ever undertaken. It is estimated to cost about $3 billion over a period of fifteen years and is hoped to have completed its sequencing work by the year 2005.

The task facing the project is colossal. The human genome contains an estimated 100,000 genes, which encode information in DNA for making a human individual in the first place, and for maintaining the individual in his/her daily life. These genes are distributed among twenty-three pairs of chromosomes – twenty-two pairs of so-called autosomes and a pair of sex chromosomes. Each chromosome contains a long DNA molecule combined with various protein molecules. Human genes are made up of four bases – A, C, G and T – known as nucleotides, linked together like beads on a string along long twisted strands of DNA. The chromosomes of each human cell contain a total of about 3 billion of these nucleotides. The goal of the Human Genome Project is to determine the precise sequence of these bases in the hope of finding clues to the causes of many of the inherited diseases.

The stuff of genes is deoxyribonucleic acid (DNA), a spiral molecule resembling a ladder whose 'rungs' are built of pairs of bases. If we were to unravel the DNA of all the chromosomes in each cell in a person's body and join it end to end it would stretch to the moon and back about 8000 times. Yet, with the new techniques of molecular biology, scientists can now isolate a single gene of perhaps 1000 to 2000 bases from an amount of DNA sufficient to contain more than 6 million genes of similar size!

Even so, only 2 per cent of human genes have so far been pinpointed to specific chromosomal locations; and only a handful of some 4000 genetic diseases are understood at the molecular level. The Human Genome Project aims to locate the position of all these genes, and to read the genetic information encoded in them, including the aberrant information in disease genes. One of the disease genes already located is that mentioned in the scene – the cystic fibrosis gene – which leads to chronic and eventually lethal lung infections. Cystic fibrosis sufferers have to endure daily treatment to clear sticky mucus from their lungs, and most die in their twenties. Another gene located is that for neuro-fibromatosis, a disease that causes tumours that are sometimes so disfiguring that the Elephant Man of nineteenth-century fame was once thought to have had it. Both diseases are fairly common. In Europe one in twenty-five people carry a single copy of the cystic fibrosis gene and one in 2500 people are born with the disease. One in 3000 people are born with neurofibromatosis.

Locating the genes for these two diseases was painstaking but relatively easy – it took two years. Tracing more complex diseases will be a Herculean task, not least because many – such as one of the forms of breast cancer – are probably caused by defects in several genes. Nevertheless, even for multi-gene diseases it can only be a matter of time before the Human Genome Project unlocks their secrets.

Gene Therapy

Once the Human Genome Project has pinpointed the location and nature of genes for disease or other attributes, the way is clear for gene therapy – the replacement or repair of a missing or defective gene.

Initially, the most likely candidates for gene therapy will be single-gene disorders, because they are the most well-studied genetic diseases and the easiest to treat. There are, of course, many difficulties in applying gene therapy to a whole person. The so-called *somatic-cell* gene therapy would involve replacing a gene in the chromosomes of each one of millions of a person's host cells. Moreover, getting a gene into these cells would not inevitably mean that the gene would do its job. There is always the risk, too, that the additional gene might interfere with other processes in the normal life of the host cells.

At present, experiments on other species are helping to solve some of these technical problems. In particular, researchers are trying to persuade viruses to assist them. Certain viruses, called retroviruses, insert their own genes into the chromosomes of the cells they naturally infect. Although retroviruses cause serious diseases, including cancer, it is possible to disable them with the new techniques of molecular biology. Their disease-causing genes can be removed and a therapeutic gene 'stitched' on to the gene that enables the virus to enter cells. Experiments on other species have shown that the technique is feasible, and somatic cell gene therapy has already begun on children whose gene for an enzyme, adenosine deaminase, does not work, thus fatally weakening their immune system.

Clearly, gene therapy would be much easier – and probably much more effective – if it was carried out on a sperm, egg or very young embryo rather than a whole person. Then the target gene would need to be replaced in only one, or a few, cells. When these divide and grow into an adult, the repaired or replaced gene will have found its way into every cell in the body. Unlike matchmaking, any form of such *germ-line* gene therapy could actually rid the gene pool of the disease, thereby freeing future generations from pain and discomfort.

Several approaches to germ-line therapy are being tested. One is to cultivate and alter sperm, which can then be used for IVF or artificial insemination. Another, during IVF, is to check the genes and modify any defects at the 'pre-embryo' stage – before it is placed in the mother. Such pre-implantation diagnosis would offer women the chance of starting pregnancy knowing that their children will not inherit the disease in question, rather than finding out during the first or second three months of pregnancy. A third approach – already successful with some species – is to insert genes into the foetus. In principle, this is the same as somatic cell gene therapy – replacing genes in every cell in the body (but including the germ cells) – but is easier, because a foetus has fewer cells than an adult.

If Nathanial's mother hadn't been so impatient for a grand-child, he would probably have chosen the cystic fibrosis sufferer as a gamete partner. There would have been no reason not to – particularly if Nathanial himself did not also carry the gene. But even if he did, by 2075 germ-line therapy could have protected their child from the disease. The hope is, though, that by then the cystic fibrosis gene might even have been excised from the human gene pool altogether. If eggs had only been available, Nathanial could have had his 'number eight', cystic fibrosis carrier or not.

The current prediction is that somatic-cell gene therapy will be standard medicine soon after the turn of the century. Germ-line therapy may not be far behind – the first techniques have already been patented, albeit amidst the inevitable controversy over eugenics. Such therapeutic manipulations are little different from the Ashkenazi Jews' arrangements over Tay-Sachs but are a far cry from the nation-sized breeding programme that Galton and his colleagues had in mind or that Adolf Hitler and the Third Reich began to put in place.

The main ethical objections to modern biotechnology are dis-cussed at the end of this book. But one of the more specific focuses for debate has been where germ-line therapy should draw the line – because if the process was shown to work in extreme cases, pressure would inevitably increase for it to be used in less severe cases. Cystic fibrosis, for example, is fatal and debilitating and if the gene could be removed from the human gene pool the majority of people would probably not object. But what about diseases that are less deadly, such as those that are chronic but painful, like arthritis, for example? And is it really such a big step from asking to have undesirable genes *removed* from the embryo to requesting desirable genes – such as for tennis-playing ability – to be *added*?

Sex Ratios and Gender Selection

Nathanial and his mother had strong but opposing views on what sex the child should be – and by 2075, the chances are that people really will be able to choose. Perhaps surprisingly, though, this particular biological hurdle has proved to be one of the most difficult for biotechnology to cross.

As we have seen, each sperm and egg contains a single (haploid) set of chromosomes – twenty-two autosomes and one sex chromosome. Some sex chromosomes look like the letter X and some like the letter Y – hence they are called X- and Y-chromosomes. Eggs only ever contain an X-chromosome. Half of all sperm also carry an X-chromosome; the other half carry a Y-chromosome. The sex of a child depends on whether its mother's egg is fertilized by an X-sperm or a Y-sperm.

If an egg with its X-chromosome is fertilized by an X-sperm, the embryo is XX and will grow to be female. If the fertilizing sperm contains a Y-chromosome, the embryo is XY and will grow to be male. Since semen always seems to contain equal quantities of X- and Y-sperm, we might expect boys and girls to be born in equal numbers. Increasingly, though, there is evidence for both humans and other animals that if nature is left to its own devices, the sex ratio is not entirely random, and can depend on individual circumstances, such as people's status or occupations.

The first few children born in families of the Havasupai Indians of Arizona over the past century, for example, are more often than not boys, while later children are increasingly likely to be girls. Similar birth-order effects are well known from several human populations: Filipinos, Australian Aborigines and the Yanomamo Indians of Venezuela, to name but a few. In France and Britain, both during and immediately after the two World Wars, there was a bizarre increase in the proportion of boys among the children fathered by soldiers. Men who work as divers are more likely to produce girls – as are test pilots, clergymen, anaesthetists and men with non-Hodgkin's lymphoma. Women with hepatitis A or schizophrenia are slightly more likely to bear daughters. Finally, in Britain, Germany and the United States, 'high-status' males – defined by their inclusion in their national *Who's Who* – are more likely to sire sons. The ultimate manifestation is the sex ratio of the children of American presidents who, to date, have had ninety sons but only sixty-one daughters.

If natural selection has managed to shape people's bodies to manipulate the sex ratio of their children according to circumstance – as it has in a wide range of other species also – biotechnology ought to be able to do the same. And over the centuries there has been no shortage of claims that it can. The Greek Anaxagoras was so influential in his insistence that boys came from the right testicle and girls from the left that centuries later some sonless French aristocrats are said to have endured amputation of the left one. Most traditional advice, though, has been more user-friendly and has most often consisted of absurd diets, uncomfortable sexual positions or strategies such as facing into the sun during sex, timing intercourse with the cycle of the moon, or squirting lemon juice up the vagina before sex. None of these actually make any difference to the sex ratio of the children born, but then they are not particularly harmful either.

Modern biotechnology's attempts have involved trying to exploit the one measurable difference between X- and Y-sperm: the X-chromosome is slightly larger and contains a little more DNA than the Y chromosome, although in human sperm the difference is only 2.9 per cent. In 1973, for example, a technique was published in the journal *Nature* that has now been franchised to over fifty clinics in the United States alone. It was claimed to produce semen samples enriched in Y-sperm, because these slightly lighter sperm were better able than their heavier X-counterparts to swim through thick solutions of albumin. Later, in 1989, a technique was developed that involved incubating sperm with a dye that binds with DNA without killing the sperm. The X-chromosome, with more DNA, takes up more dye and so gives off more fluorescent light when irradiated with a laser. Those sperm that fluoresce brighter than the rest are given an electrostatic charge and deflected into a collection vessel. Finally, 'sex-associated proteins' on the surface membranes of sperm cells are different on X-sperm and Y-sperm. Antibodies that selectively bind to the X- and Y-sperm can thus be produced. When the antibodies bind to a sperm they disable it. Adding an X-antibody to a semen sample enriches its active Y-sperm content, increasing the chance of having a boy. Adding a Y-antibody increases the chance of having a girl. The proteins can also be used to immunize females against either X- or Y-sperm. In theory, this could pave the way for a woman to have prophylactic injections that dictate the sex of her next child.

Promising though all these techniques might sound in principle, the sad truth is that despite claims and counter-claims none seem to be any more effective than the folk advice that preceded them. For every practising clinic that claims success, there is an independent research laboratory that disagrees and, for the moment, couples desperate enough to pay such clinics are probably wasting their money.

By 2075, though, an infallible method *will* exist. In fact, the technology is already with us. The problem with past attempts has been that all they do is establish semen samples rich in either X- or Y-sperm which are then used for artificial insemination. Anything can happen once the sperm are swimming through the female tract. But by the time people like Nathanial are routinely using ICSI to make babies gender selection will not be a problem. Dyes already exist which can 'sex' individual sperm – painting 'male' sperm green and 'female' ones red, for example. It will then simply be a question of selecting an appropriate sperm for ICSI, and the sex of the baby can be determined with certainty.

Not all the research to pre-select babies' gender is driven by commercial exploitation of people's preferences. There are therapeutic reasons for such research, too. Defective genes on the sex chromosomes cause some diseases – disorders that are said to be sex-linked. Haemophilia – the bleeding disease – for example is due to a recessive gene found on the X-chromosome. This means that a woman only suffers the disease if she has the haemophilia gene on both her X-chromosomes. If she has it on only one she is a carrier but does not suffer. A man suffers, though, even if he inherits just one copy – because he has only one X-chromosome. A woman who has two copies of the gene, therefore, would be well advised not to have sons, to which end gender pre-selection would be invaluable. Most people, though, wish to pre-select a child's gender simply to pander to some preference, as in the scene.

So what if future couples could choose the sex of each baby? Will major imbalances in sex ratio be created that will lead to social upheaval? Most people have assumed that free and certain choice would lead to an automatic increase in the proportion of males, leading to an increase in male aggression, male homosexuality and women having multiple partners. But such scenarios are wild fantasy. Even if they became reality, they would last a very short time.

In Western industrial societies, preferences for boys and girls seem more or less evenly balanced. An American study in 1983 concluded that most women pregnant with their first child had no preference, as did a later British study. Although preferring the element of surprise in their first-born, though, nearly all women want and expect the full experience of motherhood that entails the bearing and rearing of both sexes. The result is that most people only have strong feelings over the matter if they lready have two children of the same sex.

In other societies, however, the situation can be very different. A few cultures favour females – Hungarian gypsies, for example – but most that show a strong preference favour males. A survey of settled Hungarians in the late 1970s hinted at a sex bias with males being favoured, particularly as a first-born. In Bombay during 1986 out of 8000 abortions carried out after amnio-centesis, all but one were of female foetuses. In 1989 in China, where killing girls at birth is widespread, 114 boys were regis-tered for every 100 girls.

Almost without exception, the pressures that generate such strong preferences are financial in origin. In India, the goods or money that must be paid to the groom by the bride's family under the Hindu dowry system make girls a financial liability. In Bombay, companies offering pregnant women sex tests and abortions advertise their services with the slogan, 'Invest 500 rupees now, save 50,000 rupees later'. Dowry systems are often the explanation in other cultures, too. Elsewhere – such as in China – the explanation is simply that males are better able to get jobs and/or contribute to the family income. Even when females are favoured – as among Hungarian gypsies – there is a financial explanation. Attractive and educated females can 'marry up' into the wealthier settled strata of Hungarian society whereas boys find it much more difficult.

What happens when sex pre-selection becomes widespread – as it surely will – depends largely on when it happens. If the option were available now, there would undoubtedly be an initial *worldwide* swing towards boys. If it happens in a few decades' time, the Western trend towards greater female independence may well have reached many other cultures. In which case, any swing towards males would be far less pronounced. But whenever it happens, any conspicuous imbalance in the sexes, leading for example to men being unable to find sexual partners – or even jobs – will create a self-correcting environment. It is much more likely that dowry systems and financial inequalities will change, rather than that societies will suffer major upheaval and unrest. Within at most a few decades, the pendulum would inevitably swing back until the sex ratio in the population as a whole was once more at its current level of about 106 males for every 100 females.

In the long term, there is probably little to fear from people being able to choose the sex of their baby.

The Gamete Marketing Board (GMB)

In the scene, the whole process seemed so easy, thanks to the Gamete Marketing Board (GMB) – and there is no reason why reality should be any different. The biggest headache for any such organization, of course, will be the enormous amount of bookkeeping involved, because the scope for people's gametes to be lost or mixed up will be enormous. And if, as in the case of Nathanial's father in the scene, some people end up having enormous numbers of children, the need for good genealogical records might also be very important. Or at least they will if incest is to be avoided. Whether incestuous gamete pairings would need to be avoided or not in the future is another matter (Chapter 10). In the scene it was implied that they would. Of course, technological help with bookkeeping is to hand. DNA fingerprinting, Internet technology and information storage/retrieval should all facilitate the establishment of a worldwide computerized system for gamete choice, with built-in computer controls for incest and genetic disease as described in the scene.

Maybe there will be room in the future for privately run and funded egg and sperm banks as at present. It will be essential, though, that all such satellite clinics become part of a much wider system with free exchange of information. This will both give people the widest possible choice of gamete partners and also centralize information on genealogies and genetic disease. Whether at base private clinics are initially arranged in state or national units, the ultimate need is for a worldwide organization such as the GMB.

This chapter has introduced a vision of one of the reproductive avenues of the future. The BlockBank scheme for family planning combined with a worldwide Gamete Marketing Board will be of undoubted appeal to many people – but it will not be cheap and will largely be the domain of the wealthy. The implication in the scene was that market forces would dictate how much any given person's gametes would cost.

An embryonic form of such a system is already in operation in the United States where eggs and sperm can be ordered – and surrogate mothers arranged – from web pages on the Internet. This means that no matter where in the world they live anybody with a computer terminal already has access to such a scheme. The day of the Reproduction Restaurant is not far away – when people will be able to browse through their reproductive options over a meal, a bottle of wine and a cup of coffee.

Although this chapter is entitled 'Reproduction Restaurant' and although the scene was set in just such a place, we are not yet in a position to do the institution justice. There are still many questions to answer. How will the system affect family structure and family life? What are the financial implications – particularly for child support? What range of options will be available on the menu of the Reproduction Restaurant?

These are just some of the questions addressed in Part 4. Once we have answered those, we shall be ready to paint a proper picture (Chapter 14).

Part Four

Relationships to Fear – or not to Fear

9

Commissioning a Family, Infidelity and the World's Population

SCENE 9

A Couple of Lone Parents

'How about Thursday night?' the woman asked across the breakfast table. Then, before waiting for her live-in partner to answer, she threw down her newspaper and went to the bottom of the stairs.

'Mimi,' she shouted to her clone-daughter, 'what the hell is going on up there? What's all the noise?'

'It's Hazel, Mum,' came back the answer from the twelve-year-old. 'She won't come out of the bathroom.'

'Well, use the other one.'

'But I want to use this one. It's got all my stuff in. Anyway, Harry's in the other one.'

'Well, hurry up. We've got to leave early this morning.'

'I know – so tell her,' Mimi shouted, and started banging on the bathroom door with her foot again.

The woman stalked back into the kitchen.

'Will you go and sort out your children,' she said. 'They're hogging the bathrooms. Mimi can't get in. I've got to take them to school *early* this morning.'

'I know,' he said, reluctantly putting down his newspaper and leaving the table. 'That's the tenth time you've told me already.'

She heard him walking up the stairs, shouting as he went. Despite the increase in chaos from having two extra children in the house, life was certainly easier now with an extra pair of hands – and legs – around the place.

Eventually everything went quiet and she had a few minutes of blissful peace to finish her breakfast. When her partner returned, he was dressed for work.

'I'll just go out and make sure your car starts,' he said. 'You left it out and there was a heavy frost last night. I'll scrape the windows for you, too.'

'Thanks,' she replied, gulping down the last of her coffee. Then, just as he reached the front door, she said, 'So what about Thursday night? Shall we go out?'

'OK,' he said, with no further thought. 'I guess it's our last chance until I come back from my conference. It'll make a change to go out together. Do you want me to arrange a baby-sitter, or shall we let Hazel and Harry do it?'

'Best get a baby-sitter. It's not fair on your two. I'm sure Mimi and Michael play them up when we're out of the way.'

Apart from the usual fight over who should sit where in the car, it was an uneventful drive to school despite the icy conditions on some of the more rural roads. It was an uneventful morning at work, too – three successive, hour-long, tedious committee meetings – but at least she had lunch with one of her current sex partners to look forward to.

'Thursday night,' he said. 'Let's spend Thursday night together. It's nearly a month, now.'

She reached across and placed her hand on his. 'I can't,' she said. 'I've already made arrangements for Thursday night.'

'What?'

'I'm going out.'

Briefly, he went quiet. He was by far the most jealous and possessive of her male friends. 'Who with?' he asked eventually.

'It's only my live-in partner,' she said. 'Nobody for you to be jealous of.'

He perked up again. 'Oh, well – surely you can put him off. Who would you rather spend Thursday night with? A live-in partner or a lover?'

She smiled. 'Well, to be honest, on a Thursday night I'd probably rather spend it with a live-in partner. I've got a busy day on Friday. If I spend the night with you, you'll keep me awake all night. Anyway, my period's just started. You won't want to know me by Thursday night.'

He looked crestfallen. She wasn't his only sex partner, but she was his favourite.

She patted his hand again. 'Be patient. He's going away for a fortnight on Monday. You can come and stay for a few days – help me look after the children.'

'Is he? You didn't tell me.' His pleasure at the invitation was marred by the fact that this was the first he'd heard of it.

'Didn't I? Sorry. I only found out this week. I haven't seen you since I heard.'

'That's no excuse. What's wrong with e-mail – or even the video-phone?'

'I'm never phoning *you* again. Your live-in partner was bloody rude to me last time I tried. I can't even phone you at work – what on earth were you thinking of, letting your secretary be your live-in partner – or was it the other way round? Anyway, don't go on about it. I've invited you now. I didn't have to. There are at least two other men who would jump at the chance of sharing my house and bed for a couple of weeks.'

Thursday night was a big success and proved to her all over again why she was living with him and not with any of the other men in her life. She really did enjoy his company. The only low point came when he raised the question of IVFing with her yet again.

'Why spoil things?' she said. 'You've got Hazel and Harry. I've got Mimi and Michael. It's a nice, balanced household. You look after yours financially. I look after mine. We share the hands-on care when we feel like it, and not when we don't. That was the agreement, and it works. It's certainly loads easier than when I was looking after my two on my own, and you've always said the same. Why spoil it by having a mutual?'

'But you're not getting any child support,' he said. 'Only your own. Wouldn't you like some extra – from me?'

'I don't need any extra child support, do I? If I did, I wouldn't have had a clone for my first, then bought sperm from the GMB for my second. That's no argument.'

'Are you sure that having a mutual would spoil things? I'm not convinced. I think it could actually make the household, not break it.'

'Everybody warns against mixing own and mutuals. You've seen all the studies. As soon as both parents have an interest in some children and not others, that's when all the jealousies and rivalries start. At the moment, yours *expect* me to show favouritism to mine – and mine expect you to be the same. They don't feel hurt by it. But the second they have a half-brother or -sister that gets favouritism from both of us, they'll start getting confused – and jealous. That's when all the rivalries start – and if you're unlucky, that's when the abuse starts. I don't want to risk it. There's a lot to be said for keeping parentage and live-in partnerships separate. It's worked really well for us so far. Let's not spoil it.'

She paused. It was old ground and she knew he agreed, really. Her manner softened as she continued. 'Look, if you want another baby, have one. Clone – or try one of your students,' she said mischievously. She knew it was a sackable offence to IVF with a student. 'You don't need the mother to be able to afford child tax. Waive her share. Who's going to the conference with you? I assume somebody's going to keep your bed warm?'

He told her. It was somebody she knew vaguely – but not one of his students.

'Well, there you are. Perfect. She's really bright – and pretty. Ask her. Pay her something. In fact, I bet she's banked eggs already. She'd probably be glad of the money.'

'I daresay,' he said resignedly. 'I don't know, though. It's one thing having sex with someone as young as her. It's another thing IVFing with them. You've no idea what traits are going to surface once they get out into the real world. I think I'd prefer an older gamete partner.'

'Well, think about it. I don't mind if you bring another child into the house. We can change our contract. You can pay four-sevenths of the household bills. Everything'll be fine. Anyway, let's change the subject. Tell me about your conference.'

Pedigrees and Families – Forty Years On

When asked what they fear most about the new reproductive technologies, people often mention pedigrees and relationships. They worry about the creation of families and individuals whose place in the world will be difficult to define in traditional terms. What is 'a family' when every child in the household has a different father – or mother? Who is the mother of a clone, born to a surrogate? Who is the grandmother? And so on.

Part 4 explores some of the vagaries of familial and social relationships that will be generated by reproductive technology. There are five chapters, each one focusing on a different aspect of life some time around the year 2075. One of the chapters tackles who should be paid what in the network of contributors to future reproduction. Another concentrates on how really large genetic families scattered around the world might achieve some sense of togetherness. Two other chapters deal with situations that some readers might find difficult to handle. One looks at the position of homosexuals in future society and another considers the whole question of incest and whether it will need defining afresh. This first scene, though, has dealt with something much more mundane – the role of gamete partners, live-in partners and sex partners in people's lives once the BlockBank system and child tax become a matter of normality, at least for the middle classes.

In a sense, this scene revisits the situation encountered in Chapter 2 – that of the red and green lights of fertility prediction. But whereas this scene is set around 2075, that one was set around 2035. Then, child tax and routine paternity testing had just been established, though even the middle classes still conceived via intercourse. Predatory women, armed with their fertility predictor kits, sought out likely sources of child support and tried to conceive to them. Nevertheless, the congruence of sex partner, gamete partner and live-in partner as just one person was beginning to disintegrate. Now, forty years on, they have dissociated entirely. Is this really the way of the future?

Living Together

Although the financial unit of the future is likely to be the lone-parent family (Chapter 2), it is unlikely that all such units will live separately. The situation portrayed in the scene involved two lone parent families living together for mutual assistance. As it happened, one of those families was a lone mother and the other a lone father, but any combination could have suited their respective needs. Two (or more) lone parents can benefit greatly from cohabiting, gaining from the inevitable increase in economy – of parental energy, time and watchfulness as well as space and expenses.

The two parents in the scene were both affluent. They had both waived extra child support and could probably have afforded to live separately. But for the sake of help and companionship – with maybe even an element of love – they preferred for the time being to live together. Lower down the socio-economic scale, the financial pressures to cohabit will be much greater. Low pay will ensure that lone parents need child support, but at the same time it will mean that the child support they receive will be minimal. These, therefore, will be the people with the greatest need to make financial savings by sharing accommodation.

Two lone-parent families cohabiting create a different and probably more stable situation than one in which some of the children ('mutuals' in the scene) belong genetically to both parents. Just like lone-parent families, such blended families have become an increasing feature of twentieth-century society. In the modern-day United States, for example, half of white women and one-third of black now remarry after divorce. In addition, others cohabit for varying periods of time.

Those that blend their families by having children with their second partners expose themselves to the problems we touched on in Chapter 1. A man is much more likely to abuse or kill his stepchildren than he is his genetic children. At a less extreme level, he is also more likely to show favouritism to his genetic children. And it is not only stepfathers who behave in this way. Stepmothers show all the same traits, as *Hansel and Gretel* would be the first to testify. The woman in the scene was probably right, therefore. Problems *might* have been created had the pair made a child together.

Thanks to their affluence, the couple in the scene had several reproductive options available. Although the man liked the idea of reproducing with his live-in partner, he had other alternatives.

In her turn, she had little difficulty protecting her lone-parent status. Life could be very different, though, for people at lower socio-economic levels. Their range of options will be much more limited. Forced to cohabit and to rely on the cheaper choices from the contraceptive cafeteria, they will often have little choice but to make their gamete partner at least temporarily their live-in partner. Even by the year 2075, unless there is a massive levelling of income across social strata, lower socio-economic groups are likely to be forced to retain a mixture of lone-parenthood, blended families and the occasional nuclear family. Conception via intercourse – and hence the abortions and need for infertility treatment that go with it – will continue to be features.

In short, family life at the lower end of the income scale may not be so very different from now. Family life at the higher end, though, will be very different.

The Commissioning Parent; Head of the Family

With lone parenthood as the norm and with the BB system generating freedom of family planning, there will be little pressure on the middle classes to have more than one child with any given person – unless they really want to. They will be free to make babies when they like and, to some extent, with whom they like – because of commissioning.

Several of the options open to the middle classes have been portrayed in the last three scenes. A woman or man can simply commission a child by paying for gametes from anybody else on the BB system, as Nathanial did in Scene 8. Or, if they meet somebody they would like as a gamete partner but who is not on the BB system, they can commission that other person's gametes for a one-off conception, as the man in this scene was urged to do. Again, they would probably expect to pay for them. A woman could commission a clone of herself, as the woman had done in this scene. Similarly, a man could commission a clone of himself – though to do so he would need to commission an oocyte donor and surrogate at the same time. There are many other options on the menu of the Reproduction Restaurant that we shall consider later.

As appropriate for a society of lone parents, having a family could become a very individual process. A man will be able to commission a family just as easily as a woman. Arrangements over maternity/paternity leave need be little different from now. Nor need day-care, except that this will need to be more generally available (Chapter 2). The only major change needed would be over the financial arrangements. The earliest child tax scheme outlined in Chapter 1 in which everybody supported all their genetic children becomes unfair and unworkable when children are commissioned unilaterally rather than being conceived by mutual desire. At some point in the first half of the twenty-first century, the child tax system will need to be amended.

Before a baby is made, both gamete donors will need to agree contractually to a share of the child tax for that child. Any apportionment could be agreed, but in unilateral cases the commissioning parent would usually pay 100 per cent of the child tax for that child. Non-commissioning parents, though being required to agree to the commission and the purchase and price of their gametes, would not normally expect to pay any child tax or to take any future financial responsibility. Whether they will have any *familial* contact with their non-commissioned children is another matter (Chapter 12). The same arrangement will apply if the commissioning parent clones somebody else. If they clone themselves, then naturally they are the only signatory and will take on 100 per cent of the child tax. In all cases, the full child support would be paid to the *commissioning* parent, even if it was simply their own child tax coming back to them (less the Treasury's cut!).

In the interests of uniformity, and to ease child support enforcement, *every* child except some clones will probably be registered as having a commissioning and non-commissioning parent with an agreed split in child tax between the two. In the case of a child conceived via intercourse, the commissioning parent would normally be the mother and the child tax split would normally be 50:50, with 100 per cent of the child support being paid to the mother. Any mutually agreeable arrangement, though, should be possible. Access rights, if they ever became an issue, would automatically split according to the split in child tax.

World Population in the Balance

Currently, the United Nations is predicting the world population to stabilize at around 11 billion in the year 2200. If this projection is correct, we can all sleep easy in our beds – the mid twentieth-century fear of catastrophic population growth will have been averted. The projection does assume, though, that the global decline in family size (Chapter 7) will reach the replacement level of about two genetic children per woman by around 2100. On present trends, this is what will happen – but is it going to be threatened by the mathematics of lone-parent families and by famous tennis players having hundreds of children as did Nathanial's father in Chapter 8?

Of the two, it is the mathematics of lone-parent families that is the greater threat. It is actually irrelevant that some people scatter large numbers of genetic children around the world (Chapters 7 and 12). Every child has to be *raised* by somebody. The critical factor is how many children each person will opt to raise. For the world population to remain stable, the *average* number of children in a lone-parent family will need to be one. One child *raised* per lone parent is the equivalent of each person *producing* two genetic children. But if every man and every woman commissioned two children to raise as lone parents as did the couple in the scene, this would be the equivalent of each woman giving birth to four genetic children. The world population would begin to soar.

So what is likely to happen? What would we predict from ancient urges? We have already discussed the situation for women in Chapters 7 and 8. Surrounded by minimal infant mortality, the average woman would probably opt to raise two children, and if gender pre-selection were on the menu most would probably have one of each sex. The world's population, then, will hinge on how many men opt to become lone parents. If they all do, the world population is in dire straits. If none do, the world population is safe.

The expectation would be that the world's population is safe – that the lone father in the scene would be the exception, not the rule. The number of men who would opt to become lone parents would, with any luck, be no greater than the number of women who would opt out of lone parenthood. Men's ancient urges gear them for itinerant sexual activity with minimal parental responsibility. The BB scheme alongside the GMB commissioning system and lone parenthood will allow them to do just that. Only famous men will have women queuing up to purchase their gametes, and only famous women will have men queuing up to purchase theirs.

For most men, reproduction will probably be more a case of trying to persuade women to commission children using their sperm rather than commissioning children to raise themselves. As a means of persuasion, such a man will doubtless agree to pay his share of child tax for that child. He may even offer to be a live-in partner for a while and try to convince the woman that he would be more help than hindrance. Commissioning a child will probably be a last resort for men who cannot persuade anybody to use their gametes.

The world's population prospects will be very finely balanced. If too many men begin to see appeal in lone parenthood and not enough women see *no* appeal, the population will begin to increase again. If too many women commission more than two children and not enough commission fewer than two then humankind is again in trouble.

Of course, governments could introduce legislation to control the situation but despite, or maybe because of, the twentieth-century Chinese experience of coercive family limitation, this is unlikely to be popular. And with any luck, it won't even be necessary. Ancient urges should take one look at infant mortality rates, market forces and the wider environment and as a result predispose women to *raise* an average of just under two children and men to *raise* an average of only just above zero. Genetically, both women and men will *produce* an average of two and the world population will be safe.

Infidelity – Redefinition or Oblivion?

Lone parents who commission their families will view many of the old concepts and values as increasingly quaint and irrelevant. One of the first to suffer will be the suite of concepts historically associated with long-term relationships and the nuclear family – fidelity, infidelity and promiscuity.

Traditionalists, of course, equate infidelity with adultery and define the term within the context of marriage. Towards the end of the twentieth century, though, with marriage on the decline and various forms of live-in partnerships on the increase, such a strict definition of infidelity became increasingly irrelevant. But even when infidelity couldn't exist by traditional definition, it still existed emotionally. It still *feels* like infidelity when any live-in or sexual partner has intercourse with another person whether marriage is involved or not.

We discussed why natural selection has shaped people to react so badly to their partner's infidelity in Chapter 2. Men react badly because of the risk that their partner's next child will be another man's, not theirs. Women react badly because of the risk that they might lose their practical support. And both react because of the risk that they might contract sexually transmitted diseases. This is the evolutionary logic. Consciously and emotionally, of course, this logic simply translates into a sense of hurt and betrayal.

Perhaps surprisingly, the evidence from the past is that although all these factors are important one of the most influential over people's attitudes to promiscuity is the risk of disease. A cross-cultural survey of human societies in the first half of the twentieth century showed a striking association between sexual liberation and low risk of disease on the one hand and sexual repression and high risk of disease on the other. It seems that even with the restraint of pregnancy risk, societies become sexually permissive in the absence of major sexually transmitted diseases. In a future society with no risk of either pregnancy or disease, sexual liberation would be a natural consequence of ancient urges.

So what of the future? Paternity testing, child tax, the BB scheme and hopefully medical control of all infectious diseases will mean that acts of infidelity will no longer carry any of their ancient risks. These all hasten the final dissolution of the nuclear family and the wholesale emergence of the lone-parent family. Will the concept of infidelity any longer have meaning in a world of lone parents in which sex no longer leads to reproduction?

Probably not – but ancient emotions will not disappear overnight. People are still likely to feel that certain actions are acts of infidelity. In Scene 9 and again in Scene 10 we encounter characters that feel jealous, hurt or betrayed when their live-in or sexual partners interact sexually with other people. Which acts, though, will inflict the greatest pain? Will it be, as in the past, when your sexual partner has intercourse with somebody else, even though it cannot lead to reproduction? Will it be when the genetic parent of one of your children commissions a new child with another person's gametes? Or will it be when a third party commissions your live-in or sexual partner's gametes, and they agree to the commission?

Logically, only the middle one of these three should cause pain. As long as your partner has been blocked, their intercourse will be sterile and of no consequence. And having your partner's gametes commissioned by a third party should cost you nothing and may even bring your household some extra income. But if the genetic parent of one of your children commissions another child for whom they will have to pay child tax, the result will probably be a reduction in the child support you receive. That could matter.

Emotionally, though, people are likely to respond to all such situations as if they were infidelity – and that is how they have been shown responding in various scenes. In the scenes, though, people have been portrayed as responding least to their partner's acts of intercourse with other people. For the middle classes, this is the logical outcome of future developments – but it may not be accurate. Natural selection shaped people to react badly to the image of their loved one having sex with another person. And it shaped the reaction to be largely subconscious and hormonal, rather than a cerebral calculation of costs and benefits. Even the lone parents of the future, temporarily in love with their current sexual or live-in partner, might well still suffer from their evolutionary legacy if that person has intercourse with somebody else.

We should not forget, either, that the major liberating forces would be largely the prerogative of the middle classes. Despite child support, the lower-paid will still struggle, and without affordable access to the BB scheme they will still be vulnerable to unwanted pregnancies. They will be much more dependent on their live-in partners and hence also more vulnerable to infidelity. Maybe the concept of infidelity will be reborn and redefined in future society. Maybe it will even come to be seen as a lower-class phenomenon and concept. Or maybe it will simply disappear as totally outdated and irrelevant. There is a good chance, though, that even for the middle classes ancient urges will keep the concept alive, at least emotionally, for many generations after it ceases to have any meaning logically. Sexual liberation may be total by 2075, but the old feelings of jealousy may persist throughout the twenty-first century and beyond.

A sexually liberated society that contains siblings – or rather half-siblings – that are rarely raised in the same household generates a problem. It is one pondered briefly by Nathanial in Scene 8. Siblings will become like cousins – distant and relatively unknown to each other – with a habit of popping up at the most unlikely times and places, such as in one's bed. Although *reproductive* incest will be relatively easily monitored and if necessary controlled by the Gamete Marketing Board, it will be less easy to monitor sexual incest. Even the co-copulants may not know that they are related. And if this becomes a problem for people produced via gametes, it becomcs virtually impossible for people produced via cloning. But then, the concept of incest is not the only difficulty facing clones. All family relationships become complex, as the next scene illustrates.

10

Incest, Relationships and the Law

SCENE 10

Like Mother, Like Daughter

For the fifth time in as many minutes, the agitated man shifted his position in bed and wondered how much longer he should wait before putting his plan into operation. Looking sideways at his partner sleeping soundly by his side, he knew he needn't worry about waking her. She had been sledgehammered to sleep by her regular evening cocktail of tranquillizers, anti-depressants and – forbidden but irresistible – alcohol.

For a few moments, he watched her. With the bed-covers thrown back for coolness, her naked body was lit by the full moon that illuminated the entire bedroom. Freed for the night from the irrational thoughts that haunted her waking hours, she seemed at peace. Lying flat on her back, long tousled hair splayed over her pillow, she was snoring delicately like a purring cat. The only visible sign of her inner turmoil was the way that, one after another, seemingly every muscle in her body gently twitched involuntarily, spontaneously contracting as an unfortunate side-effect of her anti-depressant. He so hoped that his 'daughter' would never go through the mental anguish that her 'mother' had suffered over the past year – but given his 'daughter's' origins he knew it was almost inevitable.

Distressed to see his partner in such a state, the man turned away and looked towards the window. The curtains and windows were wide open but the night was so stiflingly hot and humid that his naked body was still sticky with sweat. The huge house was ancient and had never been air-conditioned. He idly reached down and placed his hand on his genitals. His excitement had begun the second he had climbed into bed and even now, thirty minutes later, showed no sign of subsiding.

Getting out of bed, he walked over to the open window. He told himself he wanted the feel of the night air on his clammy body, but deep down he knew there was another reason. For a few moments, he forced himself to focus on the moon and to savour the only slightly cooler air drifting in through the window. Then, inexorably, his gaze was drawn to his left and to the window of the bedroom in the east wing. Was she awake? As he stared at what was nothing but a black and open window, he stroked the front of his erection. Suddenly appalled by his thoughts, he swung round and turned his back on the outside world, perching his buttocks on the cool window-ledge. Refocusing his attention on his sleeping, twitching partner, he did his very best to shut out the mental images that were tormenting him and which, for the past half-hour, had so powerfully fuelled his arousal.

A sound on the landing outside the open bedroom door made him hold his breath so as to listen more clearly. Was his 'daughter' also finding it difficult to sleep and on her way to the bathroom? Was this the opportunity he was looking for? But it wasn't. The sound of a torrent of urine told him the movement had been his 'son' on his nightly visit. How often had he told him to aim at the side of the bowl, not the water? His 'son' seemed to want to advertise to the world via the volume and force of his urination how well endowed he was – but the man couldn't be surprised. He had done precisely the same at eighteen, when he was his 'son's' age. The toilet flushed, the house returned to silence, and the man returned to his torment. Still slowly stroking himself, he wondered whether he would have the courage to act out his fantasy. He couldn't believe what he was contemplating, but also couldn't deny the sexual excitement his thoughts were generating.

Maybe it had always been inevitable, though he could honestly say the possibility had never occurred to him until about two years ago, when his 'daughter' was fourteen. Even now, he didn't think the possibility had occurred to his partner – yet it was so obvious, once the thought had seeded itself. If anybody was to blame for what might happen, he told himself, it was his partner. She was the one who had first suggested cloning.

When they had first met, over twenty years ago, she was just nineteen and he was twenty-one. The attraction had been immediate, mutual, and all-consuming. For at least ten years after their first meeting, they had never been happier than when in each other's company. They had discussed parenthood from the beginning. Even so, they had waited four years before starting a family.

It was the most casual of remarks that had sent them along what at the time was a minority road, reserved for accident victims younger than five. Reproductive cloning was more common now, but when they had first planned their family it was unusual, almost illicit, and still viewed with suspicion. One day, during a short stay with his parents, his partner had been looking through photographs of him as a young boy. Remarking on how angelic he looked and listening to stories about how perfect a child he had been, she had said casually how much she hoped any son of theirs would be the same.

'What you want is a clone,' his mother had joked.

They had all laughed at such a daring suggestion but the next day, as they drove home, the man's partner had announced that the idea actually appealed to her. 'Why not have one of you and one of me?' she had mused. 'That would be fair. If we have more than two, we can have the rest normally, by IVF. But wouldn't it be incredible to watch ourselves – and each other – growing up.'

He'd needed little persuasion. The idea of witnessing his own childhood via his child-clone excited him immediately. It seemed a fascinating prospect to observe at first hand the little individual that he had once been and that his parents had described to him so often. Almost as fascinating would be the experience of watching the first nineteen years of his partner's life, the changes that would eventually lead his 'daughter' to be the person, her clone-mother, he had adored the moment he met her. Some lucky man would meet the same person he'd found so irresistible, twenty years earlier.

After that drive, the only discussion was over which of them should be cloned first and whether they should tell family and friends. In the end, they decided to clone him first, but to clone her as soon as possible afterwards – and not to tell anybody.

The whole process was surprisingly easy. Only a year after he had commissioned a 'clone-son' and donated some chest cells to his newly formed cloning clinic, he was his own proud father. The whole process had fascinated him, but when it came to cloning his partner he felt totally redundant. From commissioning to birth his partner was the only one really involved. But everything worked as smoothly the second time as the first and one year and ten months after the birth of their 'son' his partner gave birth to their 'daughter'.

In truth, over the years both the man and his partner were disappointed by the exercise, or at least by the element of watching themselves growing up. Although the similarities were obvious, they could see – or thought they could see – major differences. Friends, and even parents, while occasionally remarking how much 'like his father' or 'like her mother' were the two children, still managed to find features of the son that were just like his mother and of the daughter that were just like her father. The couple would politely smile and nod, then later laugh and despair over how unobservant people could be. But even they had to admit that their children were by no means identical to them, at least not in terms of their behaviour.

They were disappointed in another way, too. Both had imagined that, because their respective clones would be genetically identical to them, they would form a relationship far closer than could be achieved with 'normal' children. This didn't happen. The man, in particular, was disappointed by his clone-son – and by his relationship with him. He was indeed a model child, just as he had been, but he found his clone-son's angelic sensitivity extremely irritating. Worst of all, however, when his clone-son passed through puberty about five years earlier, he had irritated him quite as much as he himself had irritated his own father. They often had bitter rows and long spells of not speaking to each other. Even now, just months from his clone-son leaving home to go to college, the man could not wait to be free of him for a while. There was no sign of the special relationship that he had so eagerly anticipated.

In many ways, at least in the beginning, the most rewarding aspect of their foray into cloning was not watching their own clone growing up but watching their partner's clone. Both the man and the woman thought their partner's clone more similar to their partner than their own clone was to themselves. Increasingly, they each imagined the little individual actually *was* their partner. The woman was very close to her son when he was a child, but as he moved into adulthood she became increasingly alienated from him.

The couple never did have children by the normal route. For some unidentified reason, the man's partner never conceived from routine IVF. Either the embryos failed to implant or she miscarried. Medics assured them that this had nothing to do with their decision to have their first two children by cloning, but in his partner's mind the two facts were blatantly linked. Refusing to accept that cloning was her suggestion in the first place, she blamed him for what she now saw as their 'experiment'. She blamed him for her inability to have 'proper' children. Yet she refused to use a surrogate, saying that they might as well just concentrate on the two children they had.

When she discovered three years before that he had IVFed with another woman, their relationship had disintegrated. The other woman – a colleague from work – had commissioned a child using his sperm. They hadn't even had sex, he told her. But she didn't care about his sexual exploits. She'd had her own, after all. But a gamete partner, one who managed to gestate when she no longer could – how could he?

They had nearly separated several times after she found out but then, a year ago, she had become clinically depressed. This had actually kept them together, but the truce between them, such as it was, was always uneasy and often bitter. The woman had never disowned her clone-daughter, but as her 'son' had grown more and more like his clone-father, she saw him increasingly and irrationally as a second target for blame.

Precisely the opposite had happened between the man and his 'daughter'. They had been close during her childhood, though no closer than any father and daughter. As she passed through puberty and turned into a young woman, however, he increasingly saw her as the irresistible person his partner had been when they first met rather than the broken and bitter woman she was now. Unless his memory was playing tricks on him, his 'daughter' – physically, emotionally and behaviourally – *was* his partner of twenty-one years before. He was becoming as besotted with her now as he had been with her clone-mother then. Not only that, he was sure the attraction was mutual. In recent months she had taken to sitting by his side on the settee, watching television with her head on his shoulder and hand on his thigh. Two weeks earlier, while drunk and emotional on her sixteenth birthday, she had sat on his lap while nobody else was in the room and given him a long kiss on the lips, thanking him for being such a wonderful father.

Last night, however, had provided what, for the man, had been the final straw – the event that had inflamed his mind and was now keeping him awake and aroused. It had been hot, humid and moonlit just like tonight. On an early-hours trip to the bathroom, he had walked past his daughter's room and noticed the door was slightly ajar. On his return, he had hesitated outside her door, then eventually given way to temptation and gently pushed the door further open. The moonlight picked out every detail of his daughter's naked body as she lay on top of the bed, reincarnating images of her clone-mother during the days of sexual abandonment early in their relationship. He had stood there, staring at the dark shadow between her thighs. But just as he was about to walk into the room, drawn as if by a magnet, his 'daughter' had stirred and begun to raise her head. Instinctively, he had moved away rather than be seen, but as he did so, he thought he heard her speak. Back in his own bed, he couldn't rid his mind of the image he had just seen or the licentious urge he had felt. Nor could he dismiss the words that were going round and round in his head. Was it his imagination, or had she really said 'Is that you? Come on in, I'm not asleep.'

Since fleeing from her bedroom door twenty-four hours earlier, he had alternated between congratulating himself on his self-control and flagellating himself for missing such a sexual opportunity. The word 'incest' was forever in his mind and as he looked around his office he imagined people pointing at him and chanting the word. Logically, he knew his fantasy wasn't incestuous – but deep down it felt as if it were.

As the day had worn on, though, his urge for sex with his 'daughter' increasingly gained the ascendancy over all restraining thoughts and feelings. Moreover, he became convinced that she was both expecting and wanting him to take the first step. All that evening, he had seen nuances in her smiles, eyes and words that told him tonight *must* be the night. Now, buttocks perched on the window-ledge and hand holding his unflagging penis, he knew the moment had come to act out his fantasy.

Quietly, he walked out of his bedroom and along to his 'daughter's' door. There he paused, engulfed by the chemistry of sexual arousal. He thought he heard movement inside, then convinced himself it was his imagination. Turning the handle, he gently opened the door, planning to slip into her bed before waking her with a gentle kiss and caress.

His plan failed, his 'daughter' seeing him the second he opened the door. All he could see of her was her face and legs, but that was enough because her eyes were wide open and staring straight into his. Her mouth was also wide open as she sought to stifle the sounds that were struggling to escape. But escape they did and strangled high-pitched exclamations reached his ears. A few seconds later, almost before he had time to realize what was happening, he was treated to the moonlit sight of his clone-son's naked buttocks contracting forcefully to the rhythm of insemination.

An Ancient Aversion

Most people will have read this scene with a growing sense of disgust. But for once, their enduring memory will not be of the technological horror. Any loathing for cloning will have been dwarfed by an intensity of feeling over incest. We have just watched a man lusting after his daughter, only to be beaten to his goal by his son. For many, this is the real stuff of nightmares and if this is what the future *might* hold, shouldn't we do our best to make sure it doesn't happen this way?

Intense though the reader's emotions may have been, they could be misplaced. Disgust at incest is an ancient reaction, shaped by natural selection, and is so strong, deep-rooted and emotional that it has a habit of completely bypassing the logical brain. What did we really see in this scene? We saw a man in his forties lusting after a young girl who presumably was over the age of consent in their society. Many people might disapprove – but it happens. Even if the pair had ended up having sex, it wouldn't have been illegal, because she was neither his genetic nor his adopted daughter. She was just a girl – moreover a girl with a striking resemblance to his life-long partner and with legal rights and feelings of her own.

We also saw two teenagers – both over the age of consent – having sex. Again, many people might not approve, but it happens, and it is legal. Genetically, they weren't brother and sister. They were just re-experiencing and re-enacting the same physical attraction and union as their clone-parents years earlier.

Yet, despite this input from our logical brain, the fact of the matter is that both acts still *feel* like incest. Even the man in the scene, who of all people should have known better, felt he was contemplating incest. Biologically, though, neither act *was* incestuous. But was it incestuous in legal terms? We can't answer this, because no law has yet been passed about sex between clones. This is a question for the future – one of many if legislation is to keep pace with technological developments, as legal systems around the world are already discovering. New types of people with complex pedigrees are being created, even today. The situation will become even more complex when the first human clones are produced.

In the discussion that follows, we consider first why natural selection has shaped people to have such a strong aversion to incest. Then we consider the complex relationships that are specifically associated with cloning. Third, we consider some of the problems that legal systems are already facing as a result of the new individuals that reproductive technology is creating. Finally, we project the whole question of incest and the law into the society of 2075 in which cloning, BB and gene therapy are matters of routine.

The Incest Taboo

An aversion to incest is ancient; as much a feature of other animals as humans. The reaction is programmed firmly into our genes. Every human culture so far studied prohibits mating between certain categories of people – the so-called incest taboo. Yet, despite its biological ancestry, incest has no strict biological definition. Animals simply avoid sex with relatives, and the greater the genetic relatedness, the greater the avoidance. Behaviourally, people are no different, but such sliding scales of impropriety are of little use to lawmakers. A more concrete guideline is needed. So, by law, incest is delimited as sexual intercourse between persons too closely related to marry legally, rather than by biology. Unfortunately, because marriage laws vary from society to society, acts that are incestuous in one country may not be in another.

English law defines incest as being sexual intercourse between a male and a female that the male knows to be his daughter, sister, half-sister, granddaughter or mother. First cousins can therefore have intercourse or marry and not be guilty of incest, just as they can in Japan. In America, the definition of incest varies from state to state, but in most states first-cousin marriage is considered to be incestuous.

Just as modern-day England and the different states of America differ in precisely what they define as incest, so too did pre-industrial societies. Most, however, prohibited father–daughter, brother–sister and mother–son mating – the main differences usually concerning cousins. In medieval Europe, for example, even sixth cousins were not allowed to marry. Now, *the average* degree of genetic relatedness for couples who marry in France is sixth cousin.

Although the incest taboo is universal and the biological emotions real, from time to time sub-groups within societies have felt driven to rise above their natural aversion. Usually the superior force is a perceived need to keep power, money or beliefs 'in the family'. In Egypt, brother–sister marriages were sanctioned for thousands of years among the ruling dynasties. Cleopatra was probably the most famous of the Pharaohs to marry a sibling. The Hindu-Sakta sect in India and the Mormons in Utah (until 1892) also practised brother–sister incest. These were all incestuous sub-cultures within wider societies that did not themselves practise incest. Only rarely has incest been common to a society as a whole. Exceptions were ancient Egypt (eventually) and Persia. In the Greek-Egyptian City of Arsinoe, it has been estimated that two-thirds of marriages 1800 years ago were incestuous, mainly between brother and sister.

Despite these few exceptions, it seems to be a characteristic of humans – and most other species – that although intercourse between close relatives occasionally occurs, it is normally avoided. We don't really know how often intercourse between close relatives *does* occur in modern, industrial societies. Many cases of incest must go unreported; estimates are that 75 per cent of father–daughter and even more of brother–sister incest is never reported. Against this, many of the reported cases may be unfounded. Most *reported* cases of father–daughter intercourse in humans are probably coercive, the father forcing a sexual act on a naïve or helpless victim. The same seems to be less true of brother–sister incest. Incest when both parties consent will be much more rarely reported. Very roughly, in the United States about one female in every fifty who has an elder brother will have sex with him. By comparison, about one in 150 will have sex with her father, and hardly any will have sex with their son.

About one in five cases of father–daughter incest end in pregnancy and more, perhaps half, of brother–sister.

Incest is as uncommon in birds and mammals as it is in humans. Studies have produced figures of about 1 per cent of matings in wild birds. Even in situations where incest is very difficult to avoid, such as the closely knit feral horses living in the restricted environment of the Camargue, southern France, aversive behaviour by young females manages to keep the level down to 10 per cent of all matings.

So why are humans – in common with other animals – programmed to have such an aversion to incest? And how will this aversion project into the future with its new forms of reproduction and new types of people? Something so deep-rooted must of course have an evolutionary logic, and in the case of incest this logic is straightforward. Reproductively, incest is dangerous, endowing couples who indulge with an increased chance of producing a child with a genetically transmitted disease.

We discussed the phenomenon of people carrying recessive but dangerous genes – such as those for Tay-Sachs or cystic fibrosis – when we considered gene therapy (Chapter 8). These dangerous but recessive genes are *hitchhikers*, travelling unseen through the generations until they surface in the unfortunate individual who inherits two copies of the gene. We also saw the dangers that arise when two carriers consider having a child. Then, there is a one-in-four chance that each child will inherit two copies of the gene and suffer from the disease.

Incest greatly increases the chance that two gamete partners will have exactly the same dangerous hitchhiker. We can forgo the details, but suppose a man is carrier to just one hitchhiker gene that is present in 1 per cent of the population. Any baby has a 50:50 chance of also being a carrier, but only a one in 400 hundred chance of suffering from the disease – as long, that is, as the man chooses his gamete partner from the wider society. However, if the gamete partner is a close relative – say a sister or a daughter – the chance of any child of theirs suffering the disease becomes one in eight – a massive fifty times greater risk.

The actual risks depend on how many dangerous hitchhiker genes a person carries. If he carries no such genes, there is no genetic risk from incest. If he carries many, the risk escalates. Various studies of genetic disease in children produced by father–daughter and brother–sister incest indicate that about half may end up afflicted with one genetic disorder or another. Studies of the offspring of first cousins in Japan, Brazil, France, India and Britain suggest that increase in the death of the children produced by cousins is around 4 per cent, and of uncles and nieces about 10 per cent. From simple maths, these figures indicate that most people are carrying around between one and four dangerous hitchhiker genes.

Such high risks put great pressure on natural selection to stop individual animals, including humans, from engaging in incest. Eventually, a variety of mechanisms evolved for animals as a whole, but any disgust felt while reading the scene comes from the main mechanism that people have inherited – an aversion to sex with people known throughout early life. The basic rule natural selection seems to have produced is simple: don't mate with anybody you have associated with closely during your child-hood, because he or she could be a close relative. As part of this, selection has fashioned animals' predilections for mate selection so that they find greater attraction in strangers than in familiar individuals. This preference has been demonstrated in mammals as diverse as horses, squirrels and red monkeys. Humans seem to be subconsciously programmed to follow the same rule. In a study of Israeli kibbutzim, for example, where children are raised communally, it was found that individuals never married anyone with whom they associated when young.

If people, like other animals, have evolved a range of mechanisms to avoid incest, how come it still sometimes occurs? Why do some people perform sexual acts that others could not even begin to contemplate? There are two main reasons. First, on the whole males are much more prepared to risk incest than females. This is all part of their more cavalier attitude to risky sex and the explanation is the same for incest as it was for one-night stands in Chapter 2: males have more to gain and less to lose. Second – and briefly here we are harking back to paternal uncertainty (Chapter 1) – people in the past could not tell whether they were related or not. 'Father' and 'daughter' may not be genetically related at all, if there was some suspicion that the mother had been unfaithful. Brother and sister may be only half-siblings, not full.

Natural selection will have done its best to programme people's bodies to judge relatedness and behave accordingly. And this is the final irony of incest. People *really are* programmed to avoid incestuous relationships. But if, despite this, a person's body still urges them into an apparently incestuous relationship, it is probably because they judge subconsciously that their target is not actually a close relative, in which case, union will carry little genetic risk. And *if* the subconscious is correct, the act is not incestuous, but bodies do make mistakes over such subtle judgements and real acts of incest do actually occur.

In the scene, of course, there was no real incest. Despite at least the father's innate apprehension, there was no logical or biological reason for the characters to avoid behaving as they wished. And in the future, with paternity testing and known genetic pedigrees, potentially everybody *could* know their relatedness, thanks to paternity testing. In all probability, though, only computers will know. Real people might not – or might not understand. This is because one thing is going to change dramatically in the future, and that is the *complexity* of relationships. Cloning is the main culprit, but surrogacy also plays its part.

Cloning and Relationships

In the scene, the woman was clinically depressed, the man argued frequently with his clone-son and lusted after his 'daughter', and the two teenagers engaged in a sexual relationship. At first sight, such reactions and relationships might not seem to be much of an advertisement for cloning as a means of reproduction!

None of these situations, though, were clearly attributable to cloning. In fact, even in 2000, even with non-cloned offspring, such behaviour is part of family experience for a minority of people. Nevertheless, some of the events that befell the family in the scene were clearly the result of cloning. The man and his partner had been immediately attracted to each other when they met. Little wonder, therefore, that the man found his 'daughter' attractive when she matured. Little wonder, either, that the 'son' and 'daughter' also found each other attractive. Maybe, if the man's clone-son had not been around, his 'daughter' might actually have been receptive to his sexual advances.

Clones raised a generation apart – as were father–'son', mother–'daughter' in the scene – are equivalent to identical twins raised a generation apart. As we saw in Chapter 6, they will not be identical. People who raise their own clones as children are doomed to disappointment if they expect to raise their own image and to be emotionally close to them as a result. Their clone will be raised in a different environment from themselves, and hence will develop differences, ranging from minor to major. As a result, the clone to clone-parent relationship may in fact feel little different, emotionally, from the conventional child to parent relationship. On the other hand, differences could generate relationship problems if, for example, the parent is disappointed by the less than perfect similarity.

Even the similarities might cause problems. Clone-parents may not always like their own characteristics when confronted by them in their clone-child. In the scene, the man who was so proud of having been an angelic child found the same trait in his son to be intensely irritating. Similarly, the man in the scene rebelled against his own father, and hence should not have been surprised that in his turn his clone-son rebelled against him. Such rebellion has little to do with empathy, or lack of empathy, for the parent, but is part of growing up and becoming independent. It was programmed into the boy by his genes, just as it was programmed into his clone-father by the same genes.

Inevitably, the addition of cloning to the human reproductive repertoire will have many such repercussions. Perhaps one of the most important – particularly to bureaucrats and traditionalists – will be the way it will complicate pedigrees. These will become so multi-dimensional that computers may be needed to work out even the simplest of family trees. The boy and girl in the scene illustrate just how complex these relationships will become.

The boy had two fathers and two mothers, but if his history had been only slightly different he could have had three fathers and four mothers. He had a genetic father (his paternal grandfather), a clone-father (who provided the nucleus) and a legal father (who commissioned and raised him). The last two, though, were the same man. He also had a genetic mother (his paternal grandmother), an egg-mother (who produced the oocyte; Chapter 6), a surrogate mother (who gestated and gave birth to him; Chapter 3) and a stepmother (who raised him). In the scene, his egg-, surrogate- and stepmother all happened to be the same person, but they need not have been.

The girl also had two fathers and mothers, but again she could have had more. She had a genetic father (her maternal grandfather) and a stepfather (the main character in the scene).

She also had a genetic mother (her maternal grandmother) and a woman who was her clone-, egg-, surrogate and legal mother all rolled into one. If these last four 'mothers' had all been different people, she would have had five mothers in all.

The Roman Catholic Church responded to the announcement of the first cloning of a sheep by saying that every person has the right to two biological parents, implying that cloned individuals would be deprived in having only one parent – the nucleus-donor. As we have just seen, this is not the case. If anything, the problem for a cloned individual lies with its plethora of parents, not its lack. And if parental relationships will be difficult to unravel, grandparental relationships will be a nightmare. So, too, would be the family tree of the child that could have been conceived by the girl in the scene, if we assume that both clones had not been blocked and were still capable of conceiving by intercourse.

Clearly, when cloning becomes part of the repertoire of human reproduction, it will bring with it a whole new language of family and relationships. For some people, this will seem an undesirable consequence, though perhaps not to all. Maybe, when celebrating their birthdays, the cloned children of the future, far from bemoaning their fate, will feel sorry for their twentieth-century counterparts who received presents from only two parents when they have up to seven.

Such multiple relationships will naturally require some psychological – but mainly bureaucratic – adjustment. It is not, however, beyond the human psyche. Ache children, like some other tribal societies in South America, readily adjust to the concept of having primary, secondary and even tertiary fathers, thanks to paternal uncertainty. In Christian communities, children also adjust to having parents and a number of godparents. There is little psychological difference between these situations and the multiple parenthood of cloning. The situation will just need a little getting used to.

The two scenes dealing with cloning in this book – this one and the resurrected Phoenix in Scene 6 – have avoided dwelling on the more macabre visions of cloning. They have concentrated instead on the more subtle consequences. In the long run, it will be the deluge of everyday repercussions that will be the more important to the wider future society. Doubtless there are many more such consequences, not mentioned here, that will only emerge when cloning does become part of human reproduction.

For some people, the fact that traditional relationships such as parent and grandparent will become obscured by cloning and that phenomena such as incest may become increasingly confused will be cause enough to oppose the legalization of cloning. For other people, the increased choice over how to reproduce and even the new complexity of family trees will be a source of excitement, interest and wonder.

And for the legal profession, every new complexity will help create a lucrative quagmire.

Legal Quagmires

Legal systems, whether devised by village elders or Supreme Courts, have always struggled to cope with problems that are basically biological in nature. That is why formalized incest taboos, as opposed to the biological aversions that spawned them, vary so much from culture to culture. It is also why contradictions can occur side by side. In the United States, for example, as previously mentioned, most states judge first-cousin marriages to be incestuous – unless you are Jewish and live in Rhode Island.

The new reproductive technologies are throwing up novel

problems for legal systems almost daily, and one of the most publicized examples is a recent case that began in California in 1994. After trying unsuccessfully for years to have a child – and spending $40,000 in the process – John and Luanne Buzzanca arranged to have a fertility clinic combine an egg and sperm from anonymous donors. The embryo was implanted in a surrogate who contracted with the Buzzancas to carry the baby to term. The Buzzancas chose to use anonymous donors rather than ask the surrogate to use her own egg because of an earlier case in New Jersey in which the surrogate had eventually refused to hand over the baby, saying that she was its biological mother. But the Buzzancas created their own problem. In 1995, a month before baby Jaycee was born, John Buzzanca filed for divorce and refused to provide child support.

In 1997, a Superior Court judge agreed that John Buzzanca was not by any legal definition Jaycee's father and thus could not be required to pay support. He also ruled that Luanne Buzzanca was not Jaycee's legal mother, although he noted that this particular problem might be solved if she adopted Jaycee. But from whom could she legally adopt? The anonymous genetic parents? Not according to current American law. Donors cannot be regarded as natural or legal parents. The surrogate mother and her husband? Again, not according to American law in which the surrogate mother is defined as the gestational mother, not the legal mother.

At one point, Jaycee's surrogate mother sought custody for herself, contending that she had contracted to deliver the child to a happily married couple, a description that no longer fitted the battling Buzzancas. But she later withdrew her claim. Just days after the Superior Court's ruling, a Californian court granted temporary custody of Jaycee to Luanne Buzzanca and ordered John to pay child support of $386 (£240) a week until an appeal could be heard. The court's argument was that in signing the agreement, John effectively *caused* Jaycee's conception. John

Buzzanca's defence against the order is based on Californian law, which declares that to be responsible for Jaycee he must either be her genetic or adoptive father or he must have supported her after her birth. None of these applied.

For the moment, blue-eyed Jaycee is unaware that despite having five parents – a genetic mother, a commissioning mother, and a surrogate mother, plus a genetic father and a commissioning father – she is still *legally* parentless. In media jargonese she is 'nobody's child'. Her case could well make legal waves far into the twenty-first century.

The case of Jaycee Buzzanca has highlighted how far behind the advances in reproductive technology the world's legal systems are lagging. In the future, cloning will clearly raise additional problems, but in one sense at least the technology of the future will ease the legal load, leading as it will to one-parent families and a universal commissioning system (Chapter 9). In a society in which all children will be commissioned in one way or another and child tax will be apportioned before a baby is made, no child should ever be nobody's child, except as now through parental death. They would always be the legal responsibility of the commissioning parent. This would considerably ease the legal and bureaucratic tangle surrounding the new people who are about to be created in huge numbers.

It does not, however, solve the other legal problem raised by this scene – that of incest. It is time to answer the question raised in Chapter 8 – will incest be legalized in the wake of the new technologies?

Biologically, the incest taboo exists to minimize how often children are born who have to endure genetic disease. Or, at least, this was the result of natural selection even if it wasn't the logic. *Socially*, the incest taboo exists primarily to protect vulnerable children from the coercive sexual attention of their elder relatives. However, nowadays the law on incest actually adds nothing to the laws on rape, abuse and age of consent. If incest were legalized, it would not put children at any greater or lesser risk from unwanted sexual attention from parents and siblings. We can concentrate our discussion, therefore, on the *biological* repercussions of legitimizing incest.

In Scene 8, Nathanial felt a distinction between having sex with a half-sister and choosing one as a gamete partner. His reaction came in the context of a society – or at least a social stratum – in which sex did not lead to conception. Incestuous *sexual* relationships between consenting adults ran no risk of creating children with genetic disease. In which case, should such intercourse still count legally as incest? Incestuous *gamete pairings*, though, were certain to risk such a creation – or at least they would if we allowed no role for genetic screening and gene therapy.

At present, technology only allows us to scan gametes and embryos for a limited number of genetic diseases. By 2075, though, we may assume that many more can be identified and most can be corrected or replaced. This should become a matter of routine. Before embryos are introduced into a womb, they should be scanned for genetic diseases. Or perhaps simpler, all gametes or embryos could be routinely incubated in a cocktail of gene therapy viruses that correct or replace any genetic disease genes if they are present. The disease-free embryo could then be introduced into the womb. Maybe we are talking about the year 2100 or even 2200 here, rather than 2075, but it is still within the future scope of existing technology. But if this will work for non-incestuous fusions, it will also work for incestuous. In other words, incestuous gamete pairing need no longer carry any greater risk than any other pairing.

Even if incest were legalized, it is unlikely that many fathers and daughters, or brothers or sisters, would commission children from or with each other. The ancient legacy for aversion to incest will continue to do its work at the individual level for many generations to come. But some might, and the question that society will have to ask itself is whether, with the genetic risks minimized, there is any reason to stand in their way. In Nathanial's scene, and in this, I have assumed that the subconscious aversion to incest is still strong and active. Even in 2075, legal systems – and the Gamete Marketing Board's computers – will still be programmed to weed out, prevent or even punish attempts at incest. We may wonder, though, whether this will still be the case in the year 2100.

11

Homosexuality, Nucleus Transfer and the Future

SCENE 11

Round Robin

It was a huge kitchen. Large enough for two pinewood tables – one big and one small – to fit in comfortably. The two couples were sitting at the larger table while their four children were at the small.

'Who wants coffee?' asked Jill, struggling to stand. Like Clare, sitting by her side, she was seven months pregnant.

'I'll get it,' said Paul.

'I'll help,' said Gary.

'Thanks,' said Jill, gratefully collapsing back on to the sturdy wooden chair. She'd known they would offer.

'How about you kids?' asked Gary as he started getting cups out of the cupboard. 'Do you want anything else?'

The four children looked at each other. There were two seven-year-olds, a boy and a girl, and two three-year-olds, also a boy and a girl.

'Coke,' said the elder boy.

'Mine want Coke, too,' echoed the younger, whose grammar was improving rapidly.

'You're not having any more Coke,' retorted Jill, the boys'

mother. 'You can have some water. That's why you didn't eat your dinner – because you drank too much Coke.'

'And how about a "please",' said Gary, the younger boy's father.

It was a complicated situation that the two couples had generated – and with the birth of the next two babies it was destined to become even more complicated.

The four adults – already in couples – had first met as students. Victimized by the wider society in their different ways, they had all shared a student house together in the second year of their degree course and had lived together ever since. As their circumstances had improved, their houses – and their families – had grown. Sexually, they had experimented from the very beginning, but although they often swapped partners for intercourse, in their own way the two couples stayed faithful to each other. At least, each person always *slept* with his or her main partner.

Eight years earlier, when the two oldest children were conceived, financial constraints forced them to take the cheapest option and conceive via intercourse. None of them came from wealthy backgrounds and so had not joined the BB scheme. They couldn't have guessed who had conceived to whom, but the routine paternity tests had sorted things out. It emerged that Clare's baby, a girl, had been fathered by Gary, and Jill's baby, a son, by Paul. It seemed natural to all of them, therefore – given that they could now afford at least a cheap version of IVF – that paternity should be swapped for each woman's second child. It wasn't until the previous year, though, that they could afford to commission the children they *really* wanted.

The evening followed its usual routine. Even with four adults doing the herding, bathing and reading of stories it took a couple of hours after dinner to get the children in bed and settled. The two half-sisters, who shared a bedroom, were read stories by the two women. The two half-brothers, who shared another, were read stories by the two men.

The adults then relaxed for a while, watching TV and talking about their days. There were two settees in the lounge, a couple sprawled on each. From time to time somebody would comment on seeing one of the babies give a conspicuously big squirm inside its mother's womb. The quiet calm ended, though, when Gary announced that he was feeling incredibly horny and found it necessary to expose himself to them all to prove the point.

'Put it away,' said Jill, with genuine irritation, from the other settee, all interest in men's parts having long since faded. Then, addressing Gary's partner by his side, she said, 'For goodness' sake, take him to bed and deal with him.'

'But I want to see the end of the programme,' came the reply. 'Anyway, I don't feel like sex.'

Gary responded by ceremoniously taking off all his clothes, then trying to do the same to his partner, tickling ribs in the process. 'Come to bed,' he said over the high-pitched but marginally irritated laughter. 'If you don't, I'll do it here.'

'For goodness' sake, you two,' said Jill. 'You'll wake the children. Go to bed, will you?'

In the end, Gary got his way and, still naked and aroused, he led his partner out of the room by the hand. In the bedroom, it was over very quickly once Paul had also undressed and bent over the bed. They climaxed almost simultaneously, then climbed under the duvet for a post-coital embrace.

On their way to their own bed, the two women looked in first on the children, then on the two men.

'Feeling better now?' Jill said to Gary. She was standing behind Clare in the doorway. With her arms round Clare's 'waist', she was resting her chin on her partner's shoulder.

The men grunted their feelings and told the two women to go away.

Twenty minutes later, Jill and Clare climbed inelegantly into bed together. Naked, they lay on their backs side by side. They were too heavily pregnant, now, to sleep in each other's arms and made do with just holding hands. After a few minutes of marvelling yet again at Gary's sex drive, Clare rested a hand on Jill's swollen abdomen.

'How's *our* baby doing in there?' she asked.

'Fine,' replied Jill. 'She's just started her nightly game of hockey. I'm sure she's got a whole team in there with her.' Then she reciprocated, placing a hand on Clare's swelling.

'And how's *their* baby doing in there?' she asked.

A Rich and Varied Gene Pool

'Genetic manipulation will reduce the rich variety of the human gene pool,' say reactionary technophobes. 'Genes for minority behaviour patterns – such as homosexuality – will be so regularly excised from sperm and eggs that they will disappear from the human gene pool,' they continue. So paranoid are such people that it makes them reluctant to accept the genetic basis of such behaviour in the first place.

Homosexuality *is* basically genetic. And doubtless by 2075 it *will* occasionally be excised at the request of the commissioning parent. But it won't disappear from the gene pool. In fact, it will almost certainly stay at its present level or even increase a little. It will do so because some people – those with homosexual inclinations such as Jill, Clare, Gary and Paul in the scene – *will want* the gene in their offspring. And as technology will allow them to create children together, man with man and woman with woman as we have just witnessed, their wish will be easily fulfilled.

To appreciate what is really likely to happen in the future, we first need to understand the biology of homosexuality. Then we need to understand the technology that can open the way for homosexual reproduction.

Bisexuality and Homosexuality

Any discussion of homosexuality is bedevilled by the ambiguity of the words that are used. The heterosexual majority has such a stereotyped image of homosexual men and women that situations tend to be prejudged even before discussion begins. It doesn't help that many soaps and dramas persistently portray bisexuals agonizing over whether they are homosexual *or* heterosexual – or show others agonizing on their behalf. 'He's gay – but he's married. Poor man. *Poor wife*. He must decide!' Biologically, homosexuality generates no such dilemma, as is obvious once we use the right words and look objectively at the relevant facts. Pedantic though it might seem, therefore, it is vital at the outset to define the words used in the following discussion.

Homosexual behaviour is intimate behaviour shown towards other people of the same sex. *Heterosexual behaviour* is intimate behaviour shown towards the opposite sex. *Homosexuality* is the condition that involves homosexual behaviour. A *heterosexual* is someone who in his or her lifetime only ever has intimate sexual contact with the opposite sex. A *homosexual* is someone who *sometimes* has intimate sexual contact with members of the same sex. An *exclusive homosexual* is someone who in his or her lifetime *only ever* has intimate sexual contact with members of the same sex. A *bisexual* is someone who in his or her lifetime has intimate sexual contact with both men and women. In the scene, therefore, all four homosexuals were bisexual and all four from time to time showed both heterosexual and homosexual behaviour.

At first sight, homosexuality might seem a strange phenomenon to be shaped by natural selection. It only seems strange, though, if we assume that being sexually attracted to the same sex somehow prevents a person from reproducing. The evidence is, however, that homosexuals reproduce – and that often they reproduce very successfully.

Male Homosexuality

There are a number of rarely appreciated facts about male homosexuality. For a start, such behaviour is by no means unique to humans. Almost all birds and mammals show similar behaviour. Male monkeys, for example, show the same range of homosexual behaviour as men, from mutual caressing and masturbation to anal intercourse. There are reports of male monkeys behaving exactly like Paul and Gary were implied to behave in the scene – one masturbating to ejaculation while being penetrated anally by another. Homosexual behaviour has an ancient pedigree.

In the largest and most industrial of human societies, such as Europe and the United States, about 6 per cent of men show homosexual behaviour during their lifetime, most often during adolescence. Other mammals also show their peak of homosexual activity during adolescence and early adulthood. Over 80 per cent of men who are ever going to show homosexual behaviour have done so by the time they are fifteen; 98 per cent by the time they are twenty. For most, the behaviour is intimate and genital, and often involves anal intercourse.

There is now convincing evidence from family and twin studies that homosexual behaviour is inherited genetically. Moreover, the behaviour may be linked to brain 'wiring' that is intermediate between male and female. Inheritance is more often via the mother than the father. This is because at least some of the genes involved in homosexual behaviour are on the X-chromosome, of which women have two and men have only one. Consequently, homosexuals are much more likely to have uncles and cousins with similar inclinations on their mother's side of the family than on their father's.

A genetic basis to homosexual behaviour does not mean that environment plays *no* part. Boys genetically inclined towards homosexuality may not show that inclination without the necessary experiences during adolescence. The converse could also be true, though it should be less common – boys without the genetic inclination may be seduced or forced into homosexual behaviour during childhood. Nevertheless, modern evidence suggests that homosexuals are more often than not born rather than made.

Exclusive homosexuality is, by definition, non-reproductive – but then less than 1 per cent of men are exclusive homosexuals. In contrast, bisexuality *is* reproductive and is by far the commonest context for homosexual behaviour. Roughly 5 per cent of men in industrial countries are *bisexual*. In other primates, too, homosexuality is really bisexuality. Male monkeys who have anal intercourse with other males have no lower a rate of intercourse with females than those that do not.

Two of the most conspicuous features of bisexuality in humans are a high ejaculation rate and multiple partners, both male and female. Bisexuals ejaculate twice as frequently as heterosexuals, averaging in their twenties one ejaculation every twenty-four hours. Even allowing for the greater frequency, though, bisexuals' ejaculates contain about 100 million fewer sperm than heterosexuals' ejaculates, at least during masturbation. During homosexual anal intercourse, relatively few sperm are inseminated – almost 200 million fewer than a heterosexual inseminates into a woman during vaginal intercourse.

Nearly a quarter of homosexual men have more than ten male partners in a lifetime. For some, the figure can be in the hundreds.

More importantly, though, the more male partners a bisexual man has during his lifetime, the more female partners he is also likely to have. On average, a bisexual man will inseminate more females over his lifetime than will a heterosexual man. As a result, a bisexual man is more likely to have children with different mothers than his heterosexual contemporaries. It seems that adolescent interaction with other males gives bisexuals earlier and greater experience at sexual relationships and behaviour that they can then carry over with advantage into their sexual relationships with women.

Female Homosexuality

Most of what we have just described for male homosexuals also applies to females. For example, female bisexuality is widespread among different species of mammals, birds and reptiles. It is ancient behaviour. All human societies contain bisexual women and the trait is inherited genetically. Over 80 per cent of women who show homosexual (= lesbian) behaviour are bisexual – leaving far less than 1 per cent of women in any society to be exclusively homosexual.

Sexual experience with other women seems to endow lesbians with sexual knowledge and experience that – like men – they can then carry over with advantage into any heterosexual relationships. In those heterosexual relationships, bisexual women are inseminated just as many times over their lifetime as are heterosexual women. Bisexuals are more likely to have concurrent male partners than are heterosexuals and are also more likely to have sex with two different men in a short period of time.

There are, however, a few differences between male and female homosexuality, even though most of these are simply a matter of degree. For example, across societies there are always fewer homosexual females than males, a difference that is also found in most other animals. In any given human society, there tend to be about a third to half as many female bisexuals as male. This translates in large, industrial societies into 3 per cent or so of women being homosexual compared with the 6 per cent of men.

Not only are there fewer female homosexuals than male, but on average they begin their homosexual activities a little later in life. Only 50 per cent of bisexual women have had their first homosexual experience by twenty-five; 77 per cent by thirty. Some bisexual women do not have their first lesbian experience until they are in their forties.

Another difference is that bisexual women do not have as many homosexual partners as bisexual men. Only 4 per cent exceed ten homosexual partners in their lifetime compared with 22 per cent of bisexual men. Similarly, women are more likely to have longer-lasting 'monogamous' relationships with each other than are men. A common pattern is for a female to stay in a homosexual relationship for one to three years before moving on to a heterosexual relationship. Also, older women often 'fit in' a stable homosexual relationship between successive heterosexual relationships.

In many ways, we know more about the costs and benefits of female bisexuality than we do of male. It is much easier to know how many children a woman has had at different stages in her life than it is a man – especially a man who has had many different partners. For example, we know that by the age of twenty, a bisexual woman is four times more likely than a heterosexual to have a child and that even by the age of twenty-five, she is still twice as likely. By the end of her reproductive life, though, a bisexual woman is likely to have fewer children than a heterosexual. In Britain in the 1980s, for example, one survey gave the average figures to be 1.6 children for bisexuals and 2.2 for heterosexuals. Similar figures were obtained in the United States. The trends for earlier but fewer children for bisexuals just about cancel out, giving the same overall *rate* of reproduction as for heterosexuals. This is what we should expect, as described shortly.

Taking Over

So why, if bisexuals reproduce just as successfully as heterosexuals, are only 6 per cent of men and 3 per cent of women homosexual in modern industrial societies? The answer is that although there are gains from homosexuality, there are also associated dangers. In many environments it is very much a high-gain, high-risk behaviour. One risk is the danger of being injured or killed by other members of society – homophobes with such a fear of homosexuality that it metamorphoses into aggression and victimization. Another – and probably greater risk overall – is the danger of contracting sexually transmitted diseases.

This risk is well known for males, but it also exists for females. In fact, part of the reason that bisexual women often end up with fewer children despite an earlier start than heterosexuals is that their reproductive life can be curtailed through disease. By the age of twenty, female bisexuals are more likely to have experienced genital infections. By the age of twenty-five they are more likely to exhibit abnormal cells in cervical smear tests, and by the age of thirty they are more likely to have cervical cancer. To what extent this increased risk of disease is a direct result of bisexual activity and to what extent it is simply the result of greater all-round sexual activity is not known.

When there are costs and benefits to two forms of behaviour, such as homosexuality and heterosexuality, the evolutionary outcome is often for natural selection to hold the two forms in the proportions that just balance costs and benefits. We have already seen that female homosexuals and heterosexuals have an equal reproductive *rate*. In the absence of paternity data, there are no equivalent figures for males. Whereas many male bisexuals may die without reproducing at all, some will probably have large numbers of children with different women. But with natural selection holding male homosexuality down to 6 per cent of the population, the *average* reproductive rate of male homosexuals and heterosexuals should be about equal.

Imagine, then, what might happen if the risks associated with homosexual behaviour were removed. Natural selection would no longer restrain the gene for homosexuality and its reproductive benefits would lead it soaring through the population. Everybody would be bisexual. In fact, we don't have to *imagine* such a scenario, because in many past cultures this is precisely what happened. Small isolated communities on many of the Pacific Islands were exposed to little by way of sexually transmitted diseases. Individuals susceptible to any diseases that had once existed had long since died, leaving behind only those who were immune. With small populations offering little opportunity for any mild disease organisms to mutate into new and virulent forms, there was little risk attached to any form of sexual activity – and in such societies bisexuality was the norm when they were first encountered by Western anthropologists.

Anthropologically, bisexuality is both common and socially accepted in the majority (60 per cent) of human societies. Some small island communities in Melanesia accept as normal that *all* adolescent males engage in homosexual anal intercourse. Such early phases of homosexuality sometimes blossom into short-term 'monogamous' relationships, but exclusive homosexuality is very rare. Women take for granted that their long-term partner will occasionally have sex with other men and tolerate his homosexual infidelity far more than any heterosexual infidelity. The prevailing attitude is that men should continue with their homosexual activity as long as it has no negative practical consequences for their heterosexual partner. Male homosexual behaviour is very clearly a normal part of what is essentially a bisexual scenario. Even in these societies, though, female homosexuality is still not as common as male. A rough average in these basically bisexual societies is for about 30–50 per cent of women to engage in lesbian activities.

If a reduction in disease risk has been associated in the past with populations becoming bisexual, what is likely to happen in the future? Especially when, as in the scene, technology makes it possible for homosexuals to reproduce *with each other*?

Gamete Manufacture

Natural selection has shaped homosexuality around reproduction just as much as it has shaped heterosexuality. Inevitably, therefore, homosexuals will wish to use future reproduction technology as much as anybody, and in the future they will have a variety of options, one of which will be to reproduce with each other. We have already looked at the technique of nuclear transfer in the guise of cloning (Chapter 6). Essentially, an egg's nucleus is replaced by a donor nucleus. The individual that develops then has the characteristics of the nucleus donor rather than the egg donor. This system by itself, though, does not allow two homosexuals to reproduce *together* because the transferred nucleus already has a complete (= diploid) set of chromosomes. For two people to reproduce with each other so that their child has a mixture of their individual characteristics, two nuclei, one from each person and each with only half the full chromosomal complement (= haploid), have to fuse with each other. These nuclei, though, can be obtained from either eggs or sperm.

Consider a lesbian couple, like Jill and Clare in the scene, who wish to have their own child. Both would need to donate an egg.

One of the eggs, say Jill's, would be the host. A nucleus would then be removed from one of Clare's eggs and injected by ICSI (Chapter 4) into Jill's. It might be necessary to transfer a few other bits of cell apparatus from Clare's eggs or it might even be necessary to take them from a donor sperm. But these would only be to facilitate DNA fusion and to help organize development. They wouldn't add anything further to the child's genetic makeup. To all intents and purposes, the child would be genetically Jill and Clare's. Either of them could gestate the baby – though if they wished two eggs could be 'fertilized' at once with one being placed into Jill's womb and the other into Clare's. Then both women could be pregnant at the same time and the couple could produce the equivalent of fraternal twins. The only limitation to lesbian reproduction is that two women can only produce daughters because their cells possess only X-chromosomes (Chapter 8).

The process is only a little more difficult for two men, like Gary and Paul in the scene, except that they would have to employ both an egg donor and a surrogate. This wasn't difficult for Gary and Paul who had two women only too happy to help. The nucleus could be removed from an egg as usual and replaced by a nucleus taken from one of the men's sperm, say Paul's. Then a whole sperm from Gary could be injected into the egg by ICSI. The embryo would then need to be placed into the womb of a surrogate. In the scene Clare was carrying Gary's and Paul's child and Jill was carrying her own and Clare's. Unlike lesbians, though, two men would have the option of choosing the sex of their child (Chapter 8). They could either have a daughter ('X' nucleus from both) or a son ('X' nucleus from one; 'Y' from the other).

These methods assume that everybody is fertile in the sense that they are producing either eggs or sperm. If they are not, gametes would have to be manufactured by inducing meiosis in any diploid cell from the body, then transferring the now haploid nucleus into a denucleated egg cell as just described.

In the twenty-first century, infertility will have no meaning even for people who are exclusive homosexuals and produce no gametes.

Homosexuality in the Future

By the end of the twentieth century, the demand was already growing for homosexuals to have *parental* rights equal to heterosexuals'. The claims were modest: mainly the simple right to have and to raise children in a homosexual household without prejudice from the wider society. The mechanics of having children were crude but relatively straightforward for women. A case from 1996, for example, involved a lesbian couple handing a pickled-onion jar to a male volunteer along with a request for semen. One of the women then syringed the other in a crude but effective example of DIY artificial insemination. The mechanics were more problematic for the few male couples who tried because they had to recruit a surrogate mother to have the child for them, then hand it over.

These are early examples of both the commissioning process and the cohabiting of lone-parent families that were described in Chapter 9, except that here both lone parents are the same sex. Only incidentally, they are also homosexuals. When such lone-parent child-raising platforms become the norm, there will be little reason for the outside world to know or care about the sexual orientation of the lone parents concerned.

This discussion began with the worry that gene therapy might kill off homosexuality and various other minority genes that add variety to human life. But an alternative and perhaps unexpected scenario has emerged. Evolutionary precedence suggests that if the risk of disease is removed from the sexual equation in the future – as we all hope it will be – genes for homosexuality might hi-jack the gene pool completely.

In fact, neither extreme is likely, as long as reproductive choices are left in the hands of the commissioning parents and are not directed by governments, religious institutions, or any other large and influential body, no matter how well-meaning. Such bodies *would* have the power, in true Third Reich style, to force the eradication of genes for homosexuality just as easily as they could the gene for cystic fibrosis. If individuals, however, are given the freedom to make their own reproductive and genetic decisions, atrocities like this cannot happen. Of course, some people may well request that genes for homosexuality should be excised before their commissioned embryo is placed in the gestation-mother's womb. They should be allowed to do so, not only in the interest of freedom of choice, but because others, happy with their own homosexuality, will ask for an embryo with homosexuality genes to be given preference over one without. The result of such freedom of choice will be that the frequency of minority genes – such as for homosexuality – will stay at more or less their present level in the gene pool.

The reason homosexuality swamped whole gene pools in the evolutionary past was that when not constrained by STDs homosexuals *produced more children* than heterosexuals. In the future, when the number of children a person produces will depend much less on what they learn from their sexual experiences, homosexuality is unlikely to enjoy such ascendancy. Homophobes can forget the spectre of a future dominated by homosexuals, and homosexuals can discard their paranoia about future victimization. The technology of the future will protect everybody, as long as the freedom of reproductive choice is left to individuals and not hi-jacked by dictatorial majorities.

Hopefully, this chapter has reassured all factions of the argument. En route, though, it has raised an interesting question of its own. If sexual behaviour and experience are going to have less and less influence on people's reproductive output, what are going to be the important factors? Who are going to be the reproductive successes of future generations?

12

Extended Families in the Future

SCENE 12

Keeping in Touch

The man was wheeled up to the empty reception room by his eldest son, but once they had negotiated the last small step, he took over the driving of his electric wheelchair himself.

'But are you sure it's big enough?' he asked his partner by his side. At nearly sixty, he was eleven years older than her. Their son, conceived 'by mistake' when she was still in her late teens and before they were both blocked, had just turned thirty.

'Probably,' she replied, without enthusiasm. She had little interest in their visit. This was his enterprise.

'Of course it is,' said their son. 'Even if *everybody* comes, it's only one hundred and forty-nine.'

Some thanks would have been nice, he thought. It was all very well his father dreaming and directing, but he had to do all the legwork. In his opinion, he'd done really well to find this place. Twenty hotels had said they could accommodate that number of people on his father's birthday, still months away, and he'd visited every single one of them. This was by far the most impressive, for its price.

'What do you mean, only one hundred and forty-nine?' snorted his father, indignantly. 'That's a real achievement. See if *you* can beat it.'

'Yes, Dad, I know,' said his son patiently. 'But we're talking accommodation here, not dynasties.'

As he spoke, his father wheeled himself away, heading for the far side of the room. He felt the curtains, and gave them a slight tug before turning and facing his two companions.

'I like it, well done. Have you seen a menu?'

'Don't worry, Dad. The restaurant's got a really good reputation. I checked it out.'

'That's it then. Let's book it,' the invalid said, putting his hand on his heart and pressing hard to release a trapped belch.

Sitting in the back of the car on the way home, the woman idly observed the shops and the crowded pavements. How many of these faceless people, she wondered, had never met their parents, their children or their brothers and sisters. She must be really unusual, at least among the middle classes. Not only had she known both her parents, she'd also known all four of her grandparents. Furthermore, she had given birth to – and breast-fed – three children, and raised them all to adulthood. She had even conceived one of them via intercourse. She felt proud of her family and her achievement, but she also felt just a little jealous of her partner.

That was another way in which she was unusual. She had been live-in partner to the same man for thirty years – *and* he was the father of all three of her children. How many of her contemporaries could say the same? Certainly none that she knew. But how many of them would want to say the same? Most of them had either never had a live-in partner, or had had at least three. And hardly anybody she knew had more than one child with the same man.

But then he was an unusual man. That year he had been Formula One World Champion had been the most exciting of her life. She had revelled in the kudos of being his live-in partner and companion, and mother to his son. The interviews, the travelling, the monthly exposés about their relationship – she'd loved it all.

Of course, he hadn't been faithful to her. Even while she was pregnant with their eldest son, the gaggle of young, beautiful and ever-horny women who followed him around from race to race had been too much of a temptation. She was actually the only one of his followers he had ever got pregnant, but she knew it was only a matter of time before somebody else tricked him. Eventually he agreed to join the BB scheme – against his then unbearably macho principles – on condition that she joined as well. Ironically, it was in the month after he'd had his blocking vaccination that two of his groupies managed to seduce him into impregnating them. He'd been warned that he might still ejaculate sperm for a few weeks after being vaccinated but he took no notice and ended up paying child tax for both. Even more ironically, four years after joining the BB scheme – just a year after becoming World Champion – he had the crash that ended his racing career. It lost him the use of both legs – and his penis.

She had two more children with him after that, by IVF, and the five of them had spent the last twenty years in what was as near to a nuclear family as any she knew. Before his accident, she had actually been faithful to him. Even since his accident, frustrated by his impotence, she had sought satisfaction elsewhere only occasionally, and every time with his full knowledge and encouragement. It was also he who encouraged her to pay the GMB to advertise her eggs on their web site, but nobody had ever purchased any. Maybe she should have done it earlier, when she still had her looks and was in the public eye. So there she was, nearly fifty and with just three children. And there he was with one hundred and forty-nine!

His joining the BB scheme – he had placed both sperm and breast cells with the GMB – was just before the season he won his first Grand Prix, but surprisingly few sperm were purchased until he became World Champion. She knew why – he wasn't *that* good-looking and his television persona was dreadful. He came over as hesitant, uneducated and brusque. He had far more charisma in the flesh. Once he became World Champion, though, his sperm had more takers, and one woman even purchased breast cells so as to clone him. To her amusement – because he had always been a homophobe – three of the people who purchased his sperm were men. He almost refused the sales, but she talked him into all three.

The peak of his sperm's popularity, though, came after he'd had his accident. It had been a spectacular crash, with bits of car and flames everywhere. All over the world, the end of his career was shown on television. For the next six months, the world's press followed his odyssey, reporting every time he went on or off the critical list. And when his lifelong legacy became clear, they reported with unambiguous innuendo the extent of his injuries. Eventually he was wheeled out of hospital, waving and smiling at the cameras. Over the six months he was in the media spotlight, nearly 500 women worldwide contacted the GMB about his sperm, and just under 100 eventually had his children.

They were never going to be poor. He – and later their eldest son – had good business acumen. The money from his racing days as well as the income from sales by the GMB was put to good use. In the years after the accident, his business became his life. But once it was established, and the challenge had gone, he handed more and more control over to his son, and became increasingly obsessed with his large family.

One by one he tracked them down, beginning with his clone. The information provided by the GMB was a start and many of the women – and two of the men – who had purchased his sperm were either part of the BB scheme anyway or were still living at their old address. He spent hours – months – at his home computer, sending e-mails, scouring the Net, and making video calls. He set up a family conference centre and encouraged his family, scattered across six continents, to use it, to advise and help each other, and to exchange experiences.

By the time he first made contact, most of his children were in their late teens. Some became enthusiastic members of his extended family, others never voluntarily made contact. Inevitably, many of the communications he received were requests for money, but once he made it clear that money was not part of the deal they either continued making contact for emotional reasons, or never contacted him again. In the end, there were about eighty family members who regularly, though not always frequently, joined in family debates and discussions.

The innovative idea of just once in his life having his entire family under one roof to celebrate his sixtieth birthday had occurred to him five years ago, and had increasingly become an obsession. He offered to pay for everybody's expenses, and pressured and cajoled even those who had never replied since he first contacted them. So far, he had bought just over 100 travel tickets for offspring who had agreed to come.

As the urban scenery changed to the rural surroundings in which they lived, it began to rain. The prospect of meeting all her partner's children was a nightmare, though she was fascinated at the thought of meeting his clone. Now twenty-five, he was only three years younger than his clone-father when she had first met him. She had seen him on the phone, and the resemblance was uncanny – as, of course, it should be.

As the day grew nearer, the organization became increasingly intense. With just a month to go a terrorist bomb meant that over twenty people living in Africa and the Middle East changed their minds and decided not to come. It took a hectic week of e-mails, Net searches and phone calls for her partner to find different routes, different airlines, and to persuade most of them to travel anyway. She could see he was becoming increasingly stressed and warned him to ease off; to delegate more to his son than he was doing.

Ironically, in the same way that his family was as big as it was because of a car crash, his dreamed-for family get-together was as big – and successful – as it was because of his death. Just ten days short of his birthday, he had a fatal heart attack. His eldest son arranged the funeral for his father's birthday, informed the whole family, and suggested that they come anyway. More of them made the effort and travelled to mourn his death than would probably ever have done to celebrate his birthday. Some even paid for themselves at the last minute.

Killing the Extended Family, Swamping the Gene Pool – an End to Evolution?

A genetic family is scattered across the Earth's surface in over a 100 lone parent families. A motor-racing icon has so many children that he can never meet them all. Children are created by technology, not by biology. Will the twenty-first century see the end of the extended family – or a new beginning? Can the gene pool cope with such excesses? Is the end of evolution by natural selection in sight?

Extended Families: Past, Present and Future

First came the extended family of our ancestors – children being raised by their mother, grandmother, aunts and an assortment of dubious male relatives ranging from grandfathers to fathers to uncles. This is the scheme inherited from our immediate primate ancestors, and throughout most of human evolution such an institution was the normal environment for a growing child. It still is in many cultures in many different parts of the world. People have been programmed by natural selection to look after and advise their offspring through and beyond adolescence and into adulthood, extending their influence far into those offspring's forays into parenthood.

More often than not, the ancestral extended family was patriarchal, revolving around the father's family. For example, among New Guinea highlanders a young man would wander off during adolescence, find a partner in some neighbouring or even distant tribe, and then bring her back to live in *his* extended family. Sometimes, however, the family was matriarchal, revolving around the mother's family – as, for example, among various African pastoralists and agriculturists. Although it was still the adolescent male who did the wandering, a newly formed couple would settle down with *the woman's* family.

Next, isolated in the box-homes of modern industrial society, came the nuclear family – one man, one woman and their assorted but usually shared genetic children. This is still just about the most common system in Europe, North America and various other industrial societies. The two sets of grandparents occasionally baby-sit, if they live near enough. If they don't, they have to make do with rare visits or snatched phone conversations to keep in touch with their adult offspring and grandchildren. Aunts, uncles and cousins also visit occasionally, but the extended family is a loose and scattered institution that frees the parents to pursue parenthood much as they wish.

If the future is the era of the lone-parent family, the nature of the extended family will change yet again. And at this point our traditional language for describing the situation begins to falter. The lone parent will have even greater freedom to make parental decisions, detached even from his or her children's other genetic parent(s), if there are any. The lone parent's parent(s), (half-) brothers and (half-) sisters will presumably play a role similar to now. But what about the range of non-commissioning parents, and *their* parents and brothers and sisters? Where will they fit in? Suppose a woman is lone parent to three children, each fathered by a different man who pays child tax and so has access rights. That's three men, six grandparents and any number of aunts and uncles of varying degrees of relationship. How will they keep a sense of family?

Maybe they won't. Maybe the non-commissioning parent and his or her relatives will just give up on children in other one-parent families and concentrate on their own commissioned family. It is doubtful, though, whether ancient urges will let genetic offspring go – unknown and uninfluenced – quite so easily. Like the man in the scene, many a person might try to keep in touch with their extended family, because natural selection has shaped people to want to help in the raising of children and grandchildren, if only to give advice! Unfortunately, advice based on experience is a double-edged sword.

The extended family evolved in ancestral environments that changed relatively slowly. Moreover, a child lived with one of its sets of grandparents in their home area, an environment with which they had a lifetime's familiarity. Information, abilities and techniques learned by one generation of ancestors will have been useful for decades if not centuries. It will have been largely to reap the benefits of grandparental care and advice that early human cultures will have lived in extended families.

The modern, industrial society, though, presents a quite different environment. Technological advance and the scope for travel have led to rapid social and other environmental changes over the past few generations. The situation isn't helped by people actually delaying having their first child until they are in their thirties (Chapter 7). Increasingly, people are now approaching sixty before they become grandparents. Consequently, the usefulness of their knowledge and advice to their adult offspring and grandchildren is greatly devalued. Parents become 'piggy-in-the-middle', trying to separate wheat from chaff in the shower of grandparental advice. Such devaluation of grandparental influence creates inter-generation conflict as parents reject more and more of what they see as outdated advice from their own parents – and the best way to avoid such conflict is to live well apart.

Nevertheless, grandparents – and non-commissioning parents like the man in the scene – will still have their ancient urge to keep in touch, to advise, and to try to influence the development of their genetic offspring who may indeed be scattered over six continents. Future communication technology – videophones, video conferencing – will help, as will greater accessibility of cheap air travel. But the extended family of the future will be a very different institution from that of both our ancestors and our own.

Swamping the Gene Pool

Although we may feel sorry for the man – and to a lesser extent his extended genetic family – over the difficulties experienced in trying to achieve some sense of family 'togetherness', the scene also highlights what might seem like a much more sinister aspect of the commissioning system. Unconstrained by the sheer physical labour of finding, courting and having sex with somebody before any child can be conceived, some people are clearly going to end up having enormous numbers of children. Can the gene pool cope with such major multiplication by a few people?

Of course, governments could seek to control the situation by limiting the number of children any one person could have. This has happened before – at least in the sense that sperm donor clinics put a limit on the number of offspring each donor is allowed to generate – usually around ten. An infamous transgression in the United States led to prosecution when a clinic director repeatedly used his own sperm to father significantly more than ten in a small community.

The logic behind limitation in the past was to reduce the risk of accidental incest by people who had no idea they were half-siblings. In the future, though, with GMB computers keeping track of genetic pedigrees and with gene monitoring and therapy preventing genetic diseases, there is no real reason – other than jealousy – for imposing a limit on the number of genetic offspring any one person can have.

One hundred and forty-nine children for one man might seem a lot, but in the context of a projected world population of nearly 10 billion by 2050 it really isn't many. In any case, the *range* of genes found in the human gene pool depends much more on how many people *fail* to reproduce than on how many reproduce to excess. Only if everybody who possesses a particular gene fails to reproduce will that gene actually be lost from the pool. Otherwise it will hang on and live for a chance to multiply in another generation. Of course, if any such gene is so ill-suited to the future environment that it declines to the status of hitchhiker or even disappears, then maybe its loss is nothing to be mourned anyway. There is really little to fear from the occasional excesses of a few individuals.

In fact, the whole process of differential reproductive success causing some genes to decline and some to snowball looks suspiciously like the natural selection of the past. But isn't evolution supposed to be dead in the twenty-first century?

Evolution: Alive, Well and Shaping the Twenty-second Century

Will modern technology be the death of evolution? Obviously not – because evolution is the process of change over time and as such is eternal. But will technology kill off *evolution by natural selection* – by placing the fate of genes in the hands of people rather than the natural environment?

Of course, we could be pedantic here and say that people are a 'natural' part of the environment, so that whatever forms of selection people perpetrate they are still 'natural'. Such a stance is certainly defensible, but it isn't terribly useful or interesting. So *is* anything going to be different about selection in the future?

There is one development that will to some extent be new. Some genes – the dangerous hitchhiker genes – were very resistant to natural selection. If evolution could have got rid of them completely then it would have done. But it couldn't, so it did the next best thing and masked them instead. Now they can go – eliminated from the gene pool by gene therapy (Chapter 8) – and good riddance. This process is still selection – natural selection even – but the mechanism is more powerful. 'People power' will add a veneer of force to natural selection that it never had before, giving it that extra bit of leverage to finish jobs that it couldn't quite manage on its own.

The rest of the future's selection process, though, will be unadulterated, old-fashioned natural selection. As individuals go about their business of commissioning children from some people but not others, the evolutionary process will wend its traditional way. Genes will jostle for survival and supremacy just as they have always done. The environment will change, so new genes will win an edge while ancient favourites lose theirs, but the mechanism will be the same as ever. For as long as future environments cause different genes to rise and fall in the gene pool in response to 'market' forces, natural selection will be in operation. This ancestral force will as much shape the humans of the twenty-second century as it did those who saw the dawn of the twenty-first. Future technology may occasionally give natural selection a helping hand, but most of the time evolution will proceed in its own eternal way.

So, if natural selection will continue to operate through the twenty-first century, which genes will it choose to favour?

And the Winners Are . . .

According to the rules of natural selection, genes are winners if they increase in proportion in the gene pool and losers if they decrease, or disappear altogether. It's not difficult, then, to spot some of the genes that are about to hit a winning streak as the human reproductive environment moves into twenty-first-century mode. Any gene, for example, that makes sperm, eggs or nuclei particularly robust during IVF or ICSI will do very well in the twenty-first century and could surge through the gene pool. So, too, might genes that aid implantation when an embryo is placed in the womb.

Sadly, it's also not that difficult to spot some of the losers. Genes for seduction technique, sperm production, vaginal lubrication, menstruation, penile erection and the other instruments of reproduction via intercourse are doomed to decline. The man in the scene, for example, never needed to seduce any of the women – or men – who commissioned his children. He even achieved most of his reproductive success *after* losing his penis.

Not all the changes to ancient structures will be negative, though. Cervical mucus, for example, need no longer compromise between being penetrable to sperm and impenetrable to viruses and bacteria (Chapter 3). No longer will it need to allow sperm through at all. So any gene that makes the mucus impenetrable to everything will increase in the gene pool, and women's health will benefit accordingly.

Genes can only be victorious if the people who contain them produce more genetic descendants than is average for the population. Although in principle our main interest is in ascendant and descendant genes, it's sometimes much more fun to think in terms of which people will reproduce most. It seems inevitable that the emergence of the GMB and the BB system will allow some people to gain enormous reproductive success. The biggest successes will be not so much those who commission lots of children – because they will always be limited by the cost of raising those children – but those who are the most sought-after non-commissioning parents, as in the scene.

Any genetic qualities that make a person more likely to be chosen as a gamete partner will be favoured by natural selection. Just what those qualities might be are difficult to predict, and for a while the physical qualities of attractiveness are likely to be a factor (Chapter 8). Increasingly, though, the ability to gain the public eye – and the genes that allow a person to do so – will become important. Presumably, therefore, genes for competitiveness, single-mindedness, cunning, outright ability and maybe intelligence – in other words, anything that helps a person to achieve fame and status – will be favoured.

In principle, there is nothing new in this. From the best hunters amongst our ancestors to rock stars and politicians at the present day, *men* of fame and status have always had access to more women and produced more 'love-children' than their less famous or powerful brethren. The most famous case was Ishmail the Bloodthirsty, an ex-emperor of Morocco. Maybe not all of the 888 children he claimed to have sired were actually his, but even give or take the occasional infidelity on the part of his harem members, it's a fairly safe assumption that his reproductive success was well above average. Even the motor-racing icon in the scene didn't manage to match Ishmail, but with 149 children he did reasonably well – and at least he *knew* that all the children were his!

In this context, one of the most appealing aspects of future reproduction is that no longer will there be an inequality between the sexes. The ancient legacy was that men, with their astronomical number of sperm, were limited only by opportunity whereas women were limited by their bodies (Chapter 2). In the future, though, as long as a woman can bank enough eggs with the GMB she can have as many children as demand allows. And if she also banks cells for culture and the *manufacture* of gametes (Chapter 11), there is no limit to her number of children. So far, the highest number of children ever recorded for a woman is sixty-nine over twenty-seven pregnancies. The most famous women of the future will doubtless beat that figure many times over, as we shall see in the next scene.

13

Future of the Oldest Profession

SCENE 13

Making a Living

'Applaud' said the board held high by the short, tubby man bedecked with headphones standing at the front. And as the TV host bounded down the steps, cordless microphone clutched firmly in his right hand, that's just what the audience did. After the host had taken several bows, the cue board changed to 'Stop', and the applause died down.

As was his style, the host got straight to the point. 'Welcome to my show,' he said. 'This week we're asking the question – making money out of motherhood: is it every woman's right or is it just a new form of prostitution? On the stage, we have four women who, in very different ways, all manage to make money out of motherhood . . .' he gestured to the four women who smiled and nodded while the audience clapped on cue ' . . . and in front of them – or do I mean confronting them – we have members of the public. Go on, give yourselves a round of applause too.'

When the noise died down, he climbed on to the podium to kneel by the side of by far the most glamorous of the four women. Thrusting his microphone towards her with one hand, he mopped his forehead in mock sexual excitement with the other.

'Now, Wendy. You need no introduction – the promiscuous sex

siren at the heart of everybody's favourite soap. But recently, you've made the news for another reason. Tell us about it.'

'Well, Justin. One of the tabloids published a league table of celebrities based on how much money they earn from selling gametes . . .'

'You mean eggs and sperm,' he interrupted. Most people who watched his show would think a gamete was something foreign you found in a butcher's shop.

'That's right, Justin, eggs and sperm – and also how many children they have. And in this country, they reckoned I was the top woman.'

'And what were the figures?'

'They *said* I'd earned a million from my eggs and that around the world I have two hundred children.'

The audience went completely quiet.

'Were they right?' pressed the host.

'I'm not telling you,' said Wendy, her voice rising an octave.

'But it's a lot?' Justin added. He didn't press her. A condition of her appearing was that he didn't try to get the *real* figures out of her.

'It's a lot,' she agreed sheepishly.

The host stood up and walked behind, dragging a hand across her bare shoulders and neck as he went.

'Right, thank you, Wendy. Back to you later.' He knelt by the side of the next woman. She was older and much less shapely than the actress. 'Now, Victoria – tell us how you exploit motherhood.'

'I don't *exploit* motherhood,' she said gruffly, in a surprisingly coarse voice, the result of years of chain-smoking. 'I meet a need. Obviously, people like Wendy don't have all two hundred of their children themselves. In fact, most of them don't have *any* of their children themselves. A surrogate does all the hard work. I run an agency for surrogate mothers. We've got five thousand names on our list now – and we're not the only agency around by any means. So any man lucky enough – or maybe I mean wealthy enough – to purchase one of Wendy's eggs contacts us to find a woman to gestate the baby for him. The most successful of my women are pregnant every two years. It's very lucrative. They probably earn more than you do, Justin.'

Victoria laughed at her mischief, and so did some of the audience, but Justin didn't.

'Thank you for that, Victoria,' he said. 'Back to you, later, too.'

He moved alongside a smart and attractive woman in her late forties. Of the four, she definitely came top for poise and elegance.

'Now, Jane. You don't have two hundred children, do you?'

'That's right, Justin,' she recited (they had rehearsed the opening lines several times during the day). 'But I do have ten – and I've given birth to them all and looked after them all.'

'So you're the old-fashioned type, then? Don't tell me – you've got a husband at home as well.'

Jane laughed. 'No, of course not – I'm a lone parent like everybody else. My children have got ten different fathers, all of them rich. I offered them sex for six months and motherhood at the end of it in exchange for 100 per cent child tax. I made no secret of it. I collect child support, it's as simple as that – and do very well out of it, too, thank you.'

There were a few boos and hisses from men in the audience.

'Yes, well, we've all heard of women like you,' said the host, departing from the script as he moved on. 'Anyway, thank you, Jane. Speak to you later.'

He knelt by the side of the last woman. She was also young and shapely, but her striking face was cold with dark eyes. She was dressed in black.

'Now then, Vanessa. You make your money slightly differently, don't you? Tell us about it – but remember there may be youngsters watching.'

'Well, Justin. It's like this.' Her face lit up and transformed her whole demeanour as she began to speak. 'I'm also employed by an agency. So many children are conceived by IVF these days, by adding sperm to eggs without the man and woman ever meeting. Usually, sperm have been collected months or even years in advance and just frozen.'

Vanessa paused. Whoever had written her script had a low impression of the audience to think she needed to say all this. The viewers would need to have been raised in a cave not to know what happened during IVF. Taking a breath, she continued. 'All over the world, there are men in rooms doing the necessary to produce sperm for freezing. It's a very lonely business. So my agency thought they would gain an edge over the competition by employing people like me to help men collect their sperm – to make it more fun.'

'So what do you do?'

'Whatever the customer wants. Most, though, just want to have sex. I wear a special internal condom that collects the sperm for handing over to the clinic and eventually to the GMB.'

'GMB?'

'Come on, Justin, everybody knows the GMB – the Gamete Marketing Board.'

'And how much do you charge for this service?'

'The agency charges a hundred for a standard thirty-minute slot, and I get eighty per cent'

'And how many hours a week do you work?

'Nine to five, Monday to Friday.'

'So you earn about five thousand a week.'

'I certainly do – and that's without overtime,' she said smugly.

Justin jumped up and moved to the front of the stage. 'So there we are. Women making money out of motherhood. Is it every woman's right or is it just a new form of prostitution? We'll see what our audience thinks immediately after the break. Don't go away, now.'

Pimps and Prostitutes . . . Come in from the Cold

If the future unfolds as portrayed in this book, reproduction will become big business. Already there are clinics all over the world making huge amounts of money from artificial insemination and IVF. So, too, are the lawyers that meander their way through intransigent cases – such as the Jaycee Buzzanca affair described in Chapter 10. Nobody has questioned the rights of the respected professions – medicine and law – to make money out of other people's reproduction. Yet the slightest hint that *a woman* might make money from hands-on participation at the sharp end of the process is viewed with immediate outrage or suspicion. Of course, all the women in the scene *were* either prostitutes or pimps, involved in selling the female body for profit. So how should we feel about this?

Biologically, a prostitute is an individual who offers other individuals sexual access in exchange for one or more resources. In humans, the resource sought and offered is usually money, but it might just as well be food, shelter or protection. In the pursuit of their profession, traditional prostitutes brave the cold of the streets, the heat of massage parlours or the smoke of bordellos. They risk exploitation, attack, murder and sexually transmitted diseases, and end every night's work with a cervix teeming with the sperm of a dozen or more men. Yet prostitutes are an almost universal feature of human societies, both now and in the past.

Prostitutes wouldn't exist if men didn't visit them and pay them. But they do. In ancient Greece and Rome, almost all men inseminated a prostitute at some time in their lives. In the United States in the 1940s, 69 per cent of men had inseminated a prostitute at least once and 15 per cent did so on a regular basis. In the UK in the 1990s, 10 per cent of men had paid for sex at least once in their lives by the time they were about fifty.

Anthropologically, only 4 per cent of societies claim not to contain prostitutes. The remainder acknowledges their presence. It is difficult, however, even in these societies, to estimate what proportion of women engage in prostitution at some time during their lives. Estimates of the number of overt prostitutes active at any one time range from less than 1 per cent of women in Britain in the late 1980s to about 25 per cent of women in Addis Ababa, Ethiopia, in 1974. All such estimates, however, are unreliable and are probably underestimates. More women than this will *sometimes* engage in prostitution.

Part of the problem lies in the lack of a precise yardstick over what really is prostitution. Overt, promiscuous sex in exchange for money is, in many ways, simply the least ambiguous exchange of sexual access for resources or 'gifts'. Throughout human history and culture, there are examples of men giving the woman (or her family) a 'gift' around the time of first intercourse without the woman being classed as a prostitute. Often the exchange is ritualized, as during marriage ceremonies. Even on the wedding night, the woman or her family may demand money before intercourse is allowed to take place. This is no more than socially sanctioned prostitution. Clearly, there are degrees of prostitution. In principle it is difficult to know where to draw the line between a traditional prostitute exchanging insemination for money and the average woman in a long-term relationship exchanging insemination for support, protection and 'gifts'.

The antiquity of prostitution extends further than it being the 'oldest profession'. It is an ancient legacy, many other animals also exchanging sex for resources. Perhaps the extreme is the female empid fly that trades sex for food. For a male empid to be given the opportunity of mating, he first has to catch a gnat from a swarm. When he finds a waiting female empid he offers her his prize. While she eats her meal, he is allowed to mate. Once he has gone the female waits for the next male to bring her food and to mate. In some species, females are so successful as prostitutes that they never need to find food for themselves.

The technology of the future will spawn a multitude of ways in which women can make money from reproduction, and the scene has highlighted four. Until there are artificial wombs and perhaps artificial eggs a man will always need a woman to donate an egg and to provide a womb before he can reproduce. Even if he doesn't intend to fertilize the egg with his sperm, he will at least need it for nucleus transfer. In contrast, a woman *can* manage without a man, such as during lesbian fertilization or self-cloning.

Of course, men can be prostitutes just as well as women. Most of the ways in which women earned money in the scene would also be open to male counterparts. Wendy made money from selling her gametes. A man can do that, too – as we saw with the racing driver (Scene 12). Similarly, if he was particularly sought after and was prepared to be a lone parent a man could emulate Jane in the scene and collect child support. Finally, Vanessa's agency probably employed a few males to help homosexual men get more fun out of donating their sperm for storage. What a man cannot do, however, is become a surrogate mother. This is one domain over which women have the monopoly. Coincidentally, it is also the domain that has created the biggest resistance to payment.

One of the most puzzling aspects of the surrogacy issue at the end of the twentieth century is the view held by moralists in many countries – the United States is an exception – that surrogate mothers should not receive payment for their services, other than, at most, modest expenses. The principle seems to be that surrogacy arrangements are acceptable as long as no one makes a profit from them. Payment, it seems, will detract from the surrogate's generosity and altogether cast doubt on her reliability and motives. No such restrictions, of course, are placed on the obstetricians and lawyers involved in the process, whose motives or reliability are not treated as suspect just because they receive a high income. But there seems to be a widely held opinion that a woman agreeing to be a surrogate mother should be expected to undergo nine months of pregnancy, preceded by lengthy negotiations and physical interference, followed by childbirth and handing over the baby all for pocket money.

Surrogacy is often denounced as an exploitation of the surrogate mother, but the idea that the exploitation is increased if she receives payment is surely paradoxical. Many surrogate mothers are indeed poor; but if they received a substantial payment this would surely be a way of alleviating their poverty. Would it be so terrible for a woman actually to earn a living from helping other women to have babies?

Of course, if the surrogate is paid for her services, the tables might turn and the surrogate could appear to be exploiting the commissioning parent(s), taking advantage of their infertility. As things stand at present, commercial exploitation of the infertile is apparently a worthy aim for scientists, physicians and lawyers who enjoy high salaries and even public accolades for their services. But similar exploitation by women for reproductive purposes attracts strong moral disapproval. Contrast the £60,000 the Australian couple in the Theresa McLaughlin saga (Chapter 3) handed over to infertility specialists – with nothing to show for it – with the total of £15,000 that Theresa McLaughlin herself received for carrying and giving birth to *five* children. Who are the pimps and who are the prostitutes – and who is exploiting whom?

We can only hope that by 2075, moral prejudice will have been set aside so that everybody involved in the reproduction process will be rewarded fairly for their contribution. Market forces would seem to be a more impartial arbiter than the average ethical committee.

Part Five

Time-Warps

14

Reproduction Past Menopause
– And Beyond

SCENE 14
The Three Musketeers

Stunned, Melissa put down the phone and sat back. Joan was dead.

She reached for her handbag and fumbled for cigarettes and lighter. To hell with the 'No Smoking in Offices' rule. Her hand was shaking – partly from her hangover but mainly from the shock. She imagined – she could almost feel – tears welling up in her eyes. But no tears came. The ability to cry seemed to have deserted her years ago, along with all other outward signs of emotional softness. But this was different. Joan deserved her tears. She ought to be able to cry for her friend, but she couldn't.

Her cigarette was half-smoked when her secretary came into the room and looked at her disapprovingly.

'Don't say a word,' said Melissa.

'But the meeting,' said her secretary, looking at the clock on the wall. 'The agenda.' She waved a sheet of paper in the air.

'Sod the meeting. Sod the agenda. I've just had a shock. Get me a coffee – a strong, black coffee with lots of brandy.'

'But it's only ten o'clock.'

'*Now*,' ordered Melissa.

She took a long draw on her cigarette, inhaling deeply. Last night had been just like old times – her, Abi and Joan sitting round a table, drinking bottle after bottle, chain-smoking, slandering old friends and swapping crude details of sexual escapades. At university where they first met – thirty-five short years ago – their drunken promiscuity had become legendary. They called themselves the musketeers, rampaging through their undergraduate years, sharing the musket of many a tasty man.

They had worked, too. And they had all been successful. How many such trios had ended up so influential in publishing? And it was publishing – the narcissistic award bonanza of the year – that had brought them together last night. An annual opportunity to eat, drink and bitch about the winners and the whole bloody competition.

She had shared a taxi part of the way home with Abi. It had never occurred to them that Joan would have been stupid enough to drive. She must have caught a taxi back to her office, then driven home. Silly, stupid cow, thought Melissa. A lamp-post of all things. Only fifty-five. What a waste.

Melissa was stubbing out her cigarette when her secretary returned with her laced coffee. Typically, the coffee itself was only tepid. But it was strong – and even the smell of the brandy made her feel better.

'OK, shoot,' she said as she sprang into action after downing the coffee.

Her secretary began reminding her of the agenda as Melissa moved frenetically about the office, collecting her needs.

'Don't stop. Come with me,' said Melissa as she walked into her en-suite bathroom.

Used to the situation, the secretary carried on the briefing while Melissa closed the toilet door and urinated. Without embarrassment, Melissa answered from time to time, even as she was wiping herself. Thank God she didn't have to cope with periods any more, she thought to herself as she rezipped her trousers.

'OK,' said Melissa, opening the toilet door and starting to wash her hands. 'That's the meeting. What about the rest of the day?'

'Lunch with that new author. Marketing meeting at two-thirty. TV recording studio at four. Radio interview at five-thirty.'

'And tonight?'

'Looks like an evening at home. Your first this week.'

'Oh.' Melissa hesitated, trying not to look disappointed. Much as she always thought she'd enjoy relaxing at home in the evening, she never did. 'Oh . . . good.'

She stalked back into her office, secretary a few paces behind.

'This afternoon – when I go to the TV studios,' Melissa said as she walked, without turning round, 'I want you to come with me. You can send some messages while we're in the taxi. It's always a slow journey.'

By the time Melissa climbed into the taxi at three-thirty, she had been delighted at the way the morning's meeting had swung in her favour, disappointed by the new author and bored by marketing. Her secretary climbed in behind her, complete with laptop mobile office. Melissa dictated the first letter and explained which files should be attached. Then she took a break as her secretary struggled to get the new machine to transmit.

The taxi was motionless, stuck in an endless line of traffic. Looking out of her window, Melissa focused on a lamp-post. A silly thought entered her head. Would she, one day, kill herself against this lamp-post? She wondered if Joan had ever sat like this, staring idly at the very lamp-post that one day was going to be the death of her.

Her secretary was becoming flustered. Whatever she did with the machine, all she got was an error message. Melissa became increasingly agitated. She was desperate for a cigarette, but the taxi had a large 'Thank you for Not Smoking' sign stuck on the window.

'I thought you knew how to use that thing,' she said, exasperated.

'*I* thought I knew how to use it, too. My son uses one all the time.'

'Your son! How old is he? You don't look old enough to have a son who uses one of those.'

'He's twenty,' she said with pride. Then explained, 'I was only fifteen when I had him.'

'My God,' said Melissa. 'How did you manage, financially I mean?' She was assuming the father was a contemporary.

In desperation, her secretary switched off the laptop, then re-booted. While waiting, she smiled mischievously.

'The father was one of my teachers. The idiot hadn't been blocked. He could have gone to jail for making me pregnant at fifteen. Anyway, I agreed to register the baby as OUP with the Child Support Agency in exchange . . .'

'OUP, what's OUP?'

'Oh, sorry – Of Untraceable Paternity. They only do a global DNA search if you pay them . . . anyway, I agreed in exchange for a fairly generous maintenance. To his credit, he never reneged, either. I guess he was always afraid I was going to point the finger.'

She paused, then decided to finish her story – just in case Melissa was interested. 'Then I had my daughter when I was eighteen. So I'd finished having babies even before I'd finished being a teenager. I was blocked at nineteen.'

'Do you regret it – having children so early?'

'Best thing I ever did. We're like brothers and sisters now. It's really nice being so close to them. I just can't imagine life without them.' The computer beeped and, realizing that she was in danger of sounding smug, the secretary returned to her task of trying to transmit the files. She knew Melissa had no children.

Melissa leaned forward and tapped on the cab window. 'I've got a TV interview at four,' she said. 'Do something. Take another road.'

'It's like this all over, lady. A bus has got itself stuck some-where. It's causing chaos.'

Melissa drummed her fingers on her knee for a full thirty seconds, then cracked, reaching into her bag for her cigarettes.

'Can't you read, lady?' said the driver, watching her in the mirror as she lit up. 'This is a no-smoking cab.' He pointed at the sticker.

Melissa blew her first lungful of smoke in his direction, then leaned forward so he could hear her better through the glass. 'When *you* start driving, *I'll* stop smoking,' she shouted.

She gazed at the lamp-post again, eventually relaxing as the nicotine did its work. She'd been blocked at nineteen, too – as had the other two musketeers – taking advantage of special cheap rates for students. The deal included having cells put into freeze-storage for future gamete manufacture. For twice the price, they could have had themselves generally advertised by the GMB, but none of them could afford it at the time – and somehow, none of them had ever found the time or inclination since. The thought of having children born and raised unseen had been little incentive to any of them, even if anybody had wanted their gametes.

Throw everything into a career, get to the top, *and then* commission children, they'd decided. But thirtieth birthdays had come and gone, so had fortieth, and – horror of horrors – so had fiftieth. Yet still none of them felt the time was quite right to saddle themselves with a child to care about. Of course, they wouldn't have to carry it. They could get a surrogate, or maybe even an artificial womb. They wouldn't even have to look after it. They could all afford live-in nannies. But they knew that they would need to *care about* it. A process that they were convinced would demand emotional reserves none of them felt they possessed – until now. Then, last night, Joan had announced that a month ago she had bought some sperm and given the GMB and a surrogate agency the go-ahead to produce a baby for her.

'Who did you choose for the father?' they had both asked her, fascinated.

She had teased them, trying to make them guess. But they never even got close. When she told them, they were astounded.

Disappointed – as they all had been – by every man they had ever met, she had chosen the sperm of an obscure author. She had spoken about him many times over the years, always claiming it was his work that first fired her interest in literature. Melissa and Abigail had tried reading him, but had found his work esoteric and depressing.

'But he died about twenty years before you were born,' Abigail had exclaimed.

'I know,' Joan said. 'I never thought the GMB would still have his sperm – but they have. They'd got them from one of the old sperm donor clinics when they were first formed. Isn't it fantastic? I'm going to have his baby.' She paused, looking radiant and happy – almost as if she was in the full bloom of pregnancy.

'I've made another decision too,' she continued. 'Don't try and talk me out of it. Do you remember what we used to say? . . . All for one . . .'

'And one for all,' Melissa and Abigail responded in unison. They all laughed, embarrassed that they had indeed used to say that – often – when they were students.

'Well, there you are,' Joan said, continuing to laugh. 'So, I've asked for a daughter – and I'm going to call her Melissa Abigail! My hero and my two friends, all wrapped up in my little baby.'

'Are you all right?' asked Melissa's secretary, seeing distress in Melissa's eyes. Just then the taxi began moving at last.

'Fine,' said Melissa, winding down the window and throwing out the stub of her cigarette. 'Just a silly thought.'

'Are you sure? OK, well anyway, I've got it to work. The files have been sent. Shall we do the next? I know what I'm doing now.'

Melissa shook her head. Suddenly, she was feeling very tired.

She arrived too late to do the TV recording, but was messed around so much as the studio vacillated that she was also too late for the live radio – despite the most valiant of efforts by the female taxi driver.

Stressed to breaking point, Melissa invited the secretary to help her unwind over a few drinks at an infamous bar near the radio station. But her companion couldn't delay. She was going to a concert with her two children that evening.

Melissa went to the bar anyway, sitting in the corner drinking, smoking and becoming increasingly depressed. She tried phoning Abigail – needing to talk to *someone* about Joan – but every time she was greeted by an answering machine. In the end, she took a taxi back to her town apartment, had a shower, and without eating opened a bottle of vodka. At first, she sat slumped in a chair, staring at the 'flames' of her electric fire. Then, feeling restless, she wandered around her apartment, occasionally swigging straight from the bottle. She paused in front of a photograph of herself with her parents, lifting it up so that she could see their faces more clearly.

It was twenty-five years since they had died – her mother in January, her father in December of the same year. They were both only just over sixty. Longevity did not run in her family. Maybe she had only another five years left?

She replaced the photograph and went to look out at the city skyline from her fifth-floor window, taking another swig of vodka as she arrived. What was so great about her life, anyway? A comfortable apartment in town, a wonderful house in the country, and two – or, rather, now only one – close friends. And that was about the sum of it. She wasn't going to climb any higher up the career ladder. Once at the top, where could you go? Most people she worked with hated her, and day after day she had to suffer abuse from taxi drivers, recording studios, and just about everybody she encountered. Joan had been right. *Now* was the time.

She walked over to her PC and switched on, hoping she hadn't left it too late. What if she couldn't get permission for her plan?

Maybe she would gestate the baby herself – be a woman, for once. Take maternity leave. Recharge her batteries. She was only fifty-five. Lots of women far older than her opted to go through pregnancy. While she waited for her computer to work through its programme, she lit a cigarette. Then, after logging on, she poured herself a glass of vodka before settling down to her search. She typed in 'Gamete Marketing Board' and waited.

On the GMB's home page, she selected 'stored but unadvertised' then typed in her name and social security number. Prompted for a password, she entered the one she had given at nineteen and never changed: 'A-R-A-M-I-S'. Sure enough, up came her entry, minimal though it was. There was her Name, Sex and Date and Place of Birth, then 'Eggs in store: 0; Cells in store: Yes; Permission to use after death: No'.

'Do you wish to commission?' she was asked. 'Yes.' 'Choose gamete partner?' 'Yes.' 'Name?' There were also requests for other details, if known, to limit the number of hits.

Melissa hesitated, then took a sip of vodka. Was she too late? She typed in Joan's name and date of birth. The screen flickered, then up came Joan's details. Name, Sex and Date and Place of Birth, then 'Eggs in store: 0; Cells in store: Yes; Permission to use after death: Yes'.

She sat back, staring at the screen, feeling relief, surprise and sadness all at the same time. When she had completed *her* form for the GMB all those years ago as a student she had unquestioningly ticked 'No' when asked for permission to use her cells after death. It had seemed a macabre notion at the time, to want to carry on reproducing after you'd died. She'd assumed the other two would have done the same. What premonition had led Joan to tick 'Yes'?

A message was flashing at the bottom of the screen, telling her that she had selected a gamete partner the same sex as herself and asking if she wanted to continue. When she clicked on 'Yes' there was another message, warning of a higher price for homosexual fertilization. Again she clicked to continue.

The screen cleared, then asked her to select a surrogate mother agency from a long list. She hesitated only briefly before selecting the alternative option right at the bottom – 'Self'.

The next screen summarized her choices and asked if she wanted to proceed with her order. Then the final screen thanked her for using the GMB, asked for her address, credit card details and e-mail address, and told her that she would be contacted shortly by her local clinic for her first counselling session. There – it was done.

Forgetting to log out, she meandered her way to her bedroom, bottle in one hand and glass in the other. Sitting on the edge of her bed, she stared at a picture on the wall – a large photograph of herself, Abigail and Joan during their student days. They were standing, obviously drunk and happy, with raised glasses as if toasting the photographer.

Melissa raised her glass in the direction of the picture. 'All for one,' she said, and began to cry.

Errors in Ancient Programmes

We have to assume that this scene is set slightly later than those in the previous chapters. Not that the technology is any more advanced than in previous scenes – but the characters are fifty-five, and Joan's gamete partner died twenty years before she was born, yet his sperm were available to the GMB at their formation. It doesn't really matter, but 2100 would just about allow all of these things to happen.

The scene has served two main purposes. It has tied together many of the features of the GMB system that have slowly emerged during the course of this book. In addition it has introduced two of the major features of reproduction that are likely to exist in the future – reproduction beyond menopause, and even beyond death. In fact the technology that allows women to reproduce beyond the menopause – and both men and women beyond death – is not at all futuristic. It is already with us and births have already occurred. As usual, it is only the social and legal machinery that is currently lagging behind.

The idea that such reproduction might routinely occur in the future is a challenging one that many might find distasteful. Yet the possibilities are not such a contradiction as they might at first seem. To an evolutionary biologist, menopause and death have always been important phases of *reproduction*. The modern environment, though, is gradually eroding the contributions that the two have traditionally made to people's reproductive output. It seems only fitting, then, that technology should seek to correct what are increasingly becoming errors in people's natural programming. Natural selection would have corrected the situation itself given enough generations, but technology can do it faster.

To understand these apparent contradictions, we need to appreciate how menopause and death might be viewed as reproductive strategies.

Menopause

In the extended families of the past, women took a greater interest in helping their children and grandchildren than did men – yet another legacy of paternal uncertainty. It was for this reason that women were programmed to go one step further than men in their preparation for grandparenthood. They pass through the menopause.

Menopause – the cessation of periods – marks the metamorphosis of a woman's body chemistry. The complex cycles that were so important to her childbearing life switch off, to be replaced by a body chemistry that enables her to avoid conception. The age at which the menopause occurs – between the mid-forties to mid-fifties – was fixed by evolution with exquisite precision. It was the ideal time for a woman to begin the transformation from parent to grandparent. She should by then have had the optimal number of children and the oldest should just about be having their own children – her first grandchildren.

Men do not have a menopause, at least in any physiological sense. With much less confidence in being any given child's grandparent, ancestral men gained less than women from helping. Moreover, there was no need for natural selection to extinguish a man's fertility as he aged. Further children were of little cost no matter how old he might have been. He could procreate without stress or strain for as long as young women took an interest in his genes, wealth or status. If all he had to do was inseminate a young and able woman, particularly if he could cuckold a young and able man in the process, then he could reproduce without penalty virtually to the moment he died.

Unlike today, only a small fraction of our female ancestors lived long enough to experience the menopause. The *average* life expectancy throughout much of our evolution will have been no greater than forty to fifty years. Even now the world's lowest average life expectancy at birth – in Sierra Leone – is just forty years and in at least eighteen countries in Africa, average life expectancy at birth is still fifty years or less. These, though, are averages. They do not mean that post-menopausal women do not exist in these societies, or that they did not exist in the past. And as long as *some* of those ancestors – about 1 per cent would probably be enough – survived beyond the cessation of menstruation, then natural selection would have shaped post-menopausal chemistry and behaviour.

In the modern environment, though, everything is changing.

Post-menopausal women get less and less opportunity to indulge the grandmaternal urges placed in them by natural selection, thanks to the decline of the extended family. Everybody has the potential to live much longer and to have many more years to help children and grandchildren, but the value of the help they can give in a rapidly changing world is declining (Chapter 12). Menopause is increasingly ill-adapted to the modern industrial setting. Unlike her ancestors, the modern woman could wait with advantage until her thirties or forties before starting a family (Chapter 7) and so would be better served if menopause came much later in life than fifty. As we head off into the future, the maladaptiveness of menopause might be expected to magnify.

Or will it? With egg storage, IVF and the BB system in operation, the menopause will become inconsequential rather than maladaptive. There will be no need for the Melissas of the future to view menopause as the end of anything (other than the irritation of menstruation). Menopause can be brushed aside – and the move has already begun.

Since the early 1990s, the world's press and television have shown frequent excitement over reports from the United States and Italy of women past menopause who became mothers at fifty and even sixty. The oldest so far seems to be a 62-year-old Italian woman who, in 1994, used donated eggs and her husband's sperm. All of the cases so far have been with the aid of donated eggs from younger women. In at least one case, though, the younger woman in question was the elder woman's daughter. In other words, the woman gave birth to her grandchild, just as Nathanial's mother was planning to do in Scene 8.

Success rates are currently not high for women over forty who wish to use *their own* eggs for IVF, even though in some clinics up to a quarter of patients are now over forty. IVF pregnancy rates for women over forty run at nearly 12 per cent, compared with about 50 per cent for the under-twenty-sevens and around 30 per cent for women in their thirties. Even once pregnant by

IVF, older women have a high risk of miscarriage – as many as 60 per cent of women over forty who become pregnant by IVF may lose their babies. Provided the pregnancy does not miscarry, though, once an older woman is pregnant and as long as she has scans and good care she can expect a healthy baby.

It seems fairly clear that the problem lies less with a woman's body as a gestation machine than with the egg that is fertilized. Old eggs, not old bodies, cause the difficulties. There are two main solutions, at present. One is for a woman to bank eggs or bits of ovary when she is young – the BB scheme that we have discussed at length. The other is nucleus transplant. Melissa was going to use cells donated when she was young for nucleus transfer and gamete manufacture.

It is thought that the deficiency in the eggs from older women lies not in the chromosomes but in the biological machinery that controls cell division. When nuclei from older women's eggs are transplanted into younger women's eggs from which the nuclei have been removed, only 15 per cent showed abnormalities compared with the normal 40–50 per cent. No births have yet resulted from this technique but they have from a similar technique – taking ooplasm from a young woman's egg and injecting it into the egg of an older woman.

Of course, there is more to evolved menopause than the switching-off of reproduction. Menopausal and post-menopausal women suffer a range of symptoms, from hot flushes to weakening bones, that many would prefer to avoid.

The main chemical change at menopause is that the ovaries permanently stop producing oestrogen. By giving women low doses of oestrogens, it may be possible to mask the effects of the change in ovarian function. Oestrogen therapy, however, is likely to raise the risk of womb cancer, but giving a low dose of progestogen at the same time can prevent this. Such is the logic of Hormone Replacement Therapy (HRT).

Currently, about 40 per cent of women in their fifties and

early sixties in the United States and about 33 per cent of such women in Britain now take hormones every day. Enthusiasts say HRT can stave off heart disease and bone-weakening and for many women can extend life expectancy for up to three and a half years. It might also relieve some of the symptoms of Alzheimer's disease in elderly women. On the down side, it seems to cause a small increase in the incidence of breast and some other types of cancer.

It is possible, then, that women in the future can use HRT to mask many of the symptoms of menopause and beyond. At the same time, the BB scheme or egg manipulation techniques might allow them to reproduce – either by gestating the baby themselves or via a surrogate – until well into their sixties or even seventies. In fact, there isn't even any need to stop there. Why not carry on and reproduce after death?

Reproduction after Death

To appreciate the shaping of life-span and death, we need to think in terms of natural selection acting within the context of an ancestral extended family (Chapter 12). The key principle is that beyond a certain point, a person's continued presence within that extended family ceases to benefit and begins to hinder the success of his or her descendants. The advantages of grandparental help, advice, resources and experience have always to be set against the cost of their taking up space and needing food. During early grandparenthood, when grandchildren are young, need more watching and carrying, and require relatively little food, the grandparental contribution could easily outweigh the cost of their presence. But as the grandchildren grow – in number, stature and requirements – the advantage of grandparental care becomes less and the disadvantage of their presence becomes greater. Eventually the balance tips and their continued presence becomes counter-productive.

Two ways that grandparents can delay the moment that their presence is counter-productive is to require less space (by becoming less mobile) and to require less food (by shrinking in stature). Eventually, however, having reduced their needs to a minimum, they can only cease being a drain on their descendants any further by removing themselves completely – by dying. Natural selection, therefore, has programmed people first to become frail and then to die. Women live longer than men because they contribute more to the care of grandchildren and, on average, are smaller and compete less for food. Baldly stated, women are a net plus to their descendants for longer than men and hence have been programmed to live longer. On average, women today can expect to live over four years longer than men – sixty-seven years versus sixty-three years. The female advantage is greatest in Europe – almost eight more years – and smallest in Southeast Asia, where it is just one year.

The evolutionary view of senility and death, therefore, is that hey have been scheduled by natural selection to maximize a person's aid to his or her descendants within the extended family. In the modern environment, however, natural selection's carefully crafted scheme is being undermined. Neither bodily frailty nor even death will be a barrier to reproduction for much longer. Gametes banked when young can be commissioned for repro-duction by a new generation, just as Joan commissioned her dead hero's sperm in the scene. Increasingly, people are using gametes posthumously.

In Italy, for example, a man used a surrogate mother to gestate a frozen embryo created by him and his wife who died two years before: the surrogate was his sister-in-law and the baby was successfully gestated and born. A mid-1990s survey of fertility clinics in the United States and Canada revealed that sperm samples are routinely taken from dead men at the request of their partners and families.

The law concerning the extraction and use of sperm from cadavers in the United States is unclear. Some lawyers suggest that sperm should be viewed simply as another body organ, which families can legally donate for research or transplantation. If so, sperm banks could use cadavers as donors as long as the departed have signed organ donor cards. Children have already been born after being conceived with their late father's sperm. In fact, the first court case has been heard in which such a child has sought recognition as legal daughter and heir.

In Britain, the law is more stringent but has been shaken by the case of Diane Blood, a young widow who was initially denied access to sperm taken from her husband while he lay dying from bacterial meningitis. Diane Blood wanted to be inseminated with the sperm abroad, but the British Human Fertilization and Embryology Authority said that the sample could not be released because the husband did not give his written consent, one of the provisos of an Act passed in 1990. Eventually, though, a European court upheld Diane Blood's request.

Practitioners apart, the idea of posthumous reproduction has received a mixed reaction. There can be little doubt, though, that within a few decades the commissioning of children from very old or dead icons will be commonplace. It will be very surprising if, by 2100, the process raises even an eyebrow. Moreover, within the framework of the GMB, the question of consent will be far less problematic – as in the scene.

If the evolutionary view of senility and death described earlier is correct – that it has been scheduled by natural selection to maximize a person's aid to his or her descendants within the extended family – the modern and future environments raise some interesting scenarios. Clearly medicine is gradually increasing life expectancy. More and more people can now expect to live into their sixties, seventies, eighties or beyond than could ever have done in the past. Average life expectancy at birth globally in 1995 was more than sixty-five years, an increase of about three years since 1985. It was over seventy-five years in developed countries and the world's highest, in Japan, was 79.7 years.

Is such medical achievement being helped or hindered by natural selection? It is a moot point whether a person is more likely to have their gametes commissioned while old and frail or when dead. Maybe some will attract more commissions by dying earlier rather than later – and some vice versa. But for all, the money and resources accumulated in a lifetime that previously enhanced descendants' lives are now being eaten up in keeping the ageing grandparents alive and comfortable in their extended old age. This increases pressure for an early death. So, too, does the devaluation of grandparental advice we have already discussed. On the other hand, the demise of the extended family removes much of the evolutionary pressure on people to die to schedule, which would favour greater longevity.

On balance, natural selection will probably be fairly neutral about life-spans in the future environment, which gives free rein to the medical drive to keep people alive longer and longer. Just how much and how quickly life expectancy can be extended is difficult to judge, but *Time* magazine has just predicted that by the year 2500, human life expectancy should be around 140 years.

And doubtless some people will be reproducing to the bitter end – and beyond.

Reproduction Restaurant – Babies to Go

The Reproduction Restaurant symbolizes choice, not a physical location. It refers to the tremendous and exciting range of reproductive options that will be available to future generations. But as most of those options will be available on the Internet, as many already are in the United States, doubtless some of the modern-day Internet Cafés will upgrade themselves literally into Reproduction Restaurants. The menu, though, will be equally browsable from home, as Melissa demonstrated in the scene.

It will be a few decades, though, before the full menu is available to everyone. For a while, people will need to carry an 'I am infertile' card before being allowed full Restaurant facilities. Governments, under pressure from their ethical advisers and the ill-informed fertile majority, are currently being forced to insist that reproduction technology should be limited strictly to the infertile. In Britain the influential Warnock Committee – giving a passable impression of seventeenth-century theologians in their stance over wet-nurses – pronounced surrogacy for the fertile as 'totally ethically unacceptable'. Technology should be a cure, not a commodity, ethicists say. There should be no hint of convenience or eugenics.

This is a sad and futile stance that probably will delay change for a decade – two at the most – before seeming positively quaint. Wet-nursing, then bottle-feeding, were initially an aid to women who could not breast-feed, but both ended up being an aid to those who preferred not to breast-feed. Artificial insemination was initially an aid to couples in which the man was impotent, but it ended up being an aid to women who preferred not to have intercourse, such as dedicated virgins, lesbians or simply women whose working lives left little room for meeting men. Amniocentesis was initially an aid to detect chromosomal abnormalities in time for therapeutic abortion, but it has ended up being an aid to women who prefer to choose the sex of their baby. IVF and surrogacy were first combined to aid women who could not gestate children. Who can doubt that their destiny is to aid those who would prefer somebody else to gestate their children for them?

As history shows, if there is a strong enough demand it will eventually be satisfied – especially if it is based on an ancient urge. Such demands, even from minorities, cannot be frustrated for ever. Any society that legislates against them will merely drive practitioners underground, as happened to abortion and homosexuality. There they will wait, finally resurfacing into the light of legality when the esoteric arguments against them collapse as reactionary and unsustainable. The pressure for freedom of access to the new technologies, though, could well have an unexpected springboard.

In the past, people lucky enough to be fertile have felt compassion for their less fortunate contemporaries, but if they are honest, they have also felt slightly superior. Not for them the months or years of agonizing over childlessness or the enduring of fertility treatment. Intercourse then reproduction – it was as simple as that. But the ever-growing paradox of the twentieth century has been that fertility also brings its problems. In the tight lifetime schedule of the modern career person, fertility can be an absolute bane. Especially when that person is also driven by the ancient urge to pursue, indulge and enjoy sex. Finding a mate, then going through pregnancy and parenthood, requires time, energy and emotion. While pregnant or breast-feeding, a woman can be at the mercy of hormones that lower motivation. Men, too, are at the mercy of their partner's hormones, which can often be no less distracting. The only alternative is to spend a lifetime using contraceptives that damage health, libido and even pleasure, yet still allow accidents. Unplanned pregnancies – even abortions – can disrupt even the most careful person's career plans.

Infertile people, though – at least once they know – suffer none of these problems. Infertile men or women are at no risk of unwanted pregnancies and in the very near future will be able to have children to fit exactly with their career plan – even after menopause. On top of that, they may have access to gametes unavailable to the fertile, as well as access to another woman's womb.

The time is not far away when fertile men and women – at least those with careers or aspirations – may begin to curse their fertility and demand the same rights and opportunities enjoyed by their infertile contemporaries. Then the launch of the BB scheme and formation of the GMB will not be far away, and every career person who can afford it will organize life around reproducing by IVF. When that happens, it is a very short step to the Reproduction Restaurant and a free choice for all.

So what will be on the menu of this Restaurant in 2100? One choice, naturally, will be not to visit – instead, to purchase gadgets and hormones from the contraceptive cafeteria next door. Retaining the option and unpredictability of conception by inter-course will certainly be the cheapest option, as long as unwanted pregnancies, abortion or child tax are successfully avoided and as long as the person eventually proves to be fertile. If infertile, people will have no choice, other than childlessness, but to use the Restaurant. Once there, though, everybody will be treated the same. Infertility will be meaningless.

Traditional stalwarts and the lower-paid excepted, fertile people will apply in their droves to join the BB scheme (Chapter 7). Then the Restaurant really opens up. On their first visit, men could arrange to bank haploid cells (sperm that they ejaculate, sperm from their testes, spermatids or spermatocytes) for direct use as gametes. And/or they could bank diploid cells: stem cells for injection into surrogate testes or cells from somatic tissue for culture, gamete manufacture and/or cloning. Women will have a similar choice: eggs as gametes, ovarian tissue to produce gametes, or somatic diploid cells to be used for gamete manufacture and/or cloning. They can then forget reproduction and go about their sexual lives – until the time comes to revisit the Restaurant, draw on their account, examine the menu and order a baby.

Most of the visitors to the Reproduction Restaurant will naturally be heterosexual couples in love. They will want to unite their banked gametes – original or manufactured – and to experience shared parenthood. The woman will commission the baby, the man will agree to his share of child tax, and for a while or even for life they will live together as a nuclear family. Some couples, though, will be male or female homosexuals in love. Whatever their sexual orientation, any couple could clone one or other of the pair, but undoubtedly most couples would want to be gamete partners.

Other customers will also arrive at the Restaurant as a couple, will also have chosen each other as a gamete partner, but will have no intention of living together. Most often, again, the woman would be the commissioning parent, destined to be the head of a lone-parent family, but again the man would agree to his share of child tax. There would be nothing, though, to stop a man who wants to be a lone parent from doing the commissioning.

The really expensive options on the menu will be those for lone men and women who can afford to waive child support. Or rather, who can afford to make do with having their own child tax come back to them, minus the government's cut (it is unlikely that governments will want to miss out on this source of Treasury income!) Such individuals could choose to commission a baby who is a clone of themselves – or somebody else. Or – more likely – they could commission a baby via the fusion of one of their own gametes with one from *anybody* else. The chosen gamete partner could be known to them or not, famous or not, of opposite sex or not, alive or not, and live locally or on the other side of the world. Commissioning parents could pay to have the embryo checked for diseases – or, depending on their preferences and prejudices, to be simply checked for trait genes. And, of course they could choose the baby's sex. In commissioning a baby, each lone *woman* could choose either to gestate the baby herself or to employ a surrogate. Each lone man would have to employ a surrogate. The only limitation to choice would be what the commissioning parent could afford, because different gametes and different options would cost different amounts.

Commissioning a baby, though, will be just the beginning of the process. There will need to be safeguards – for the baby, for the parent and for the Treasury. First, the non-commissioning parent would have the right to object, but not unreasonably so, and would have to justify such a stance. Second, as with adoption in the past, the commissioning parents would need to satisfy a counsellor that they were in a position, financially and emotionally, to raise the child they had commissioned. Again, though, permission should not be unreasonably withheld, nor should the counselling process be protracted. It would be the counsellor who would then report to and register the case with the child tax authority.

There would need to be safeguards, too, for children whose commissioning parent died either before they were born (as did Joan in the scene) or after, although these safeguards need be little different from now. If the non-commissioning parent was still alive and agreed to pay child tax in exchange for access rights, he or she would presumably take over the child's upbringing. If not, the onus would fall on one of the commissioning parent's genetic relatives, just as now. But also just as now, in the total absence of willing kin, responsibility would have to fall on the state. Care institutions, fostering and adoption will never become totally obsolete – although in a world where everyone is fertile, eager adoptive parents will be few and far between. Hopefully, though, in a world where most children are planned, *only* the death of commissioning parents will leave embryos and children in need of care.

There are no hidden dangers to the Reproduction Restaurant. Society will not crumble; the human species will continue along its traditionally erratic course. The babies produced will be as lovable as babies always have been. In fact, with family planning a science rather than a lottery, it is even likely that many more children will be *wanted* than now. The future should do better than the 50 per cent unplanned – 25 per cent *definitely unwanted* – that is the World Health Organization's appraisal of the current situation. With the right attitude in the wider society and the right handling by governments, the Reproduction Restaurant could be one of the most positive aspects of life in the second half of the twenty-first century.

Ah, yes . . . but is it natural?

15

Natural Dignity

SCENE 15

Back to Nature

The man woke as a large black fly landed on his face. He had been asleep for only ten minutes, but it had been a deep sleep.

Waving the fly away, he sat up, shading his eyes from the bright sun. When sleep faded and his sensibilities returned he felt pin-pricks in his back. Naked, save for a thin black belt of plaited fibres from which hung the necessities of life, he had been sleeping on a flat rock. Tiny bits of soil and gravel had embedded themselves in his skin. He brushed them away.

He sat for a few minutes relishing the sun's intense heat. From his position high up on a cliff-top he could overlook both the beach below and the sea which, sparkling in the sun, stretched uninterrupted to the horizon. He was hungry, but reluctant to leave the magical panorama. Even when he stood up, he hesitated for a moment to enjoy the sensation of the warm breeze blowing over his sweaty body. But eventually, barefooted, he walked inland for 100 yards or so.

His target was a tree that in the last week had become bedecked with heavy orange fruit. Taking a knife from his belt, he cut down the two ripest, squeezed them to confirm their juiciness, and walked back to his cliff-top perch. As he sucked on the opened fruit's flesh, juice running down his beard and dripping on to his chest and groin, he saw his family arrive on the beach below. All three of his 'wives' were there, along with five of his children. Everybody was naked, save for the obligatory black belt.

His oldest 'wife', in her fifties like himself, looked up and waved at him. Moments later, the other two women – one in her thirties, the other in her early twenties – also looked and waved. Even from this distance, the youngest woman's advanced condition with his next child was obvious. He waved back, and continued waving when his children joined in. They were shouting, but too far away to be understood.

The two younger women took off their belts and swam out into the lagoon along with the two eldest children. He toyed with the idea of making his way down the cliff path and joining them, but decided there wasn't enough time.

He heard the two sounds almost simultaneously – the distant engine and the ring of his mobile. He unclipped the tiny unit from his belt.

'Hallo,' he said.

'With you in five minutes,' came the reply, scarcely audible amongst the static.

'OK, I'll be at the pad. Are you on your own?'

'Of course.'

'OK, see you.'

He rang off, paused a few seconds, then switched to intercom mode and paged the woman on the beach.

'It's Jack,' he said. 'He'll be here in a few minutes.'

'Send him down,' she replied, 'if he's got time.'

The man made his way to the helicopter pad as the distant vibration slowly filled the skies. He had intended to collect his clothes from the shed at the edge of the clearing before the helicopter landed but was too late. He hesitated, but decided to go to meet the pilot as he was.

A young man in his mid-twenties climbed down from the cockpit and walked to meet him. The pair formed an incongruous sight, the younger man in pilot trappings, the elder with just a belt.

'Hi, Jack. Good flight?'

'Hi, Dad. Fine – but why aren't you ready? Jane did phone you, didn't she, about the meeting being brought forward?'

The man shook his head. 'By how much?'

'An hour.'

The man swore. 'Does that mean you haven't got time to go and see your birth-mother? She was looking forward to a swim with you.'

'No chance,' said Jack. 'Here, give me your mobile. I'll talk to her while you go and get dressed and get your bags.'

The man unclipped his mobile from his belt and handed it over. As he did so, his son looked him up and down and shook his head.

'One of these days, Dad, some enterprising paparazzo is going to track you down and you're going to appear like that on the front page of all the tabloids. "World Chief of GMB in naked sex holiday," it'll say. "Can we really trust this man with our eggs and sperm?" Share prices will nose-dive.'

'I don't care,' said the man. 'If I didn't get back to nature occasionally, I'd go crazy. Speak to your mother. I'll go and get dressed.'

What Is Natural?

One of the most persistent complaints as reproductive technology accelerated towards the twenty-first century was that the future seemed so unnatural. Humankind as a species was being taken further and further from its biological roots.

Yet government, religious, ethical – and public – demands for naturalness are often illogical, usually lacking in biological perspective, and totally inconsistent. Cloning is unnatural, we are told. So, too, is IVF and surrogate motherhood. Such developments should be under tight control – or banned altogether – lest they demean and erode human nature for ever.

But these developments are only the latest in a long line of assaults on 'natural' human behaviour. Earlier assaults have now become so much part of the human condition that, without pausing for thought, we can easily imagine that *they* are natural. In reality, it would be our original state not our modern state that many people would consider to be 'unnatural' – or at least undesirable. For example, humans evolved to hunt their own meat, pick their own fruit, forage for their own grain and vegetables – and to walk everywhere. Yet how many people in the industrialized world do any of these things from choice, now? Should we ban supermarkets for being unnatural and demeaning humanity?

The first unnatural assault on human *reproduction* was probably the most major. It damaged sperm production, increased the risk of genital infection, made menstrual bleeding a great inconvenience, and rendered nipples so soft and easily cracked that women often had to endure great pain to breast-feed. It probably also, indirectly, led to an increase in skin cancer. Yet nearly everywhere people have been cajoled or threatened into adopting the new technology. In the industrial world, both the law and public pressure actually enforce the *unnatural* state. It is illegal to insist on being natural, at least in public.

The technological development was, of course, clothing. The man in the scene would have been vilified not praised had he been photographed in his natural state. Shouldn't theologians and ethicists demand that the wearing of clothes be made illegal whenever the temperature rises above, say, 25°C on the grounds that clothing is unnatural and against human dignity?

A second, relatively minor, assault on natural human reproduction concerned secondary sexual characteristics – an assault that peaked in the twentieth century. Natural signals of maturity and sexual state are attacked enthusiastically by many cultures, with still no calls for legislation to ban such unnatural behaviour. Men shave off beards, women do their best to remove hair from top lips, around nipples, from legs, from their bikini line – and in many cultures even from under their arms. Razors, deodorants, perfumes and depilatory creams are big business. Some people make a great deal of money from urging and manipulating people into behaving unnaturally, and by and large they have government support and public acquiescence.

Most of these unnatural activities are irritating rather than dangerous, leading to nothing worse than an allergic reaction or razor rash. We now know, though, that underarm secretions are one of the ways that women – and perhaps men – judge the histocompatibility of potential sexual partners. The whole process is subconscious but tests show that people prefer the smell of members of the opposite sex who are histologically compatible – part of the unconscious chemistry of sexual attraction. How many unnatural couples have formed because the two people could not detect that they were incompatible, because both had manipulated their secondary sexual characters in an unnatural way? Should we ban the masking and mutilation of secondary sexual characters and insist on a return to our natural state?

A more major assault on the naturalness of human reproduction is contraception by artefact. As we saw earlier, both men and women have a wide range of 'natural' contraception mechanisms – based largely on the stress reaction – that they inherited from their primate ancestors. These mechanisms do their job subconsciously and relatively efficiently. But they do it with total disregard for the conscious life-plans of the modern man and woman. Subconscious contraception does not understand about exams and careers. It sees only windows of buoyancy and fertility.

It is not surprising, then, that the last few centuries have seen the origin, development and spread of more – and more invasive – forms of contraceptive artefacts. These are designed to accommodate family planning within the tight timetable of modern life. Sheaths, caps, IUDs, hormone pills, hormone injections and vaccinations have all been developed. Yet – perhaps not surprisingly in view of their nature – all have emotional, physiological or clinical side-effects. None are natural. Yet most societies have sanctioned their acceptability through usage.

There are still exceptions, of course. The Roman Catholic Church remains adamant that contraception is against the will of the Christian God, and the Japanese government is only just about to accept that the benefits of the oral contraceptive pill outweigh its dangers. Who can doubt, though, that one day even the Roman Catholic Church will accept the principle of contraception by artefact? It will have taken a long time but the last bastion of public support for natural contraception will have fallen. Would anybody then want to demand that we go back to nature?

An illuminating but novel twist to the natural versus unnatural debate comes with regard to masturbation and homosexuality. Both activities are ancient – shown by all species of primates and all human cultures – both are genetically orchestrated, and both are reproductive aids by design. Masturbation sheds ageing sperm for males and performs cervical housekeeping for females. Homosexuality speeds up the acquisition of sexual and relationship skills that can be transferred with advantage from homosexual encounters to heterosexual.

By any definition, both activities are natural. Yet some religions still consider both to be a sin. Many people cannot come to terms with their own urges to masturbate – let alone other people's. And in the historical but probably not the ancient past, homophobia had ascendancy over homosexuality. In fact, despite its natural status, homosexual behaviour – between consenting adults – has only recently been made legal in many countries. And in some outposts, such as the Isle of Man, the behaviour is still against the law.

Breast-feeding is natural. Not only that, it brings enormous benefits – emotional, psychological, physiological and medical – to both mother and child. In the seventeenth century, Church and State condemned women who failed to breast-feed their children but delegated the job to other women – the wet-nurses – instead. Then science produced formula milk and bottles and droves of women sanctioned the unnatural technology, preferring convenience to convention. Now, public prejudice is actually against women who breast-feed in public – or who breast-feed a child for more than a few months. How long will it be before a woman is successfully prosecuted for indecent exposure for breast-feeding in public – or for child abuse for letting a child of six or five, or even younger, suck her nipple? Where hides the supposed insistence that all things natural are good and all things unnatural are bad?

We could go on. Natural selection shaped identical twins whereas science and technology shaped clones – so we oppose clones. But natural selection shaped humans to walk whereas science and technology allows them to fly – yet we welcome planes. Natural selection shaped humans who succumbed to heart disease, but science and technology via organ transplants and artificial parts often allow them to survive – yet we welcome transplants. And natural selection shaped humans to be conceived inside their mother's body, but science and technology allows them to be conceived in a Petri dish – yet over 100,000 people have sanctioned IVF by their actions.

The public, legal and theological attitudes over what is natural and what isn't – what should be opposed and what shouldn't – is almost laughably inconsistent. The concept of naturalness is clearly no basis on which to judge our attitude towards the future. Only a person who lives naked, never shaves or uses deodorants, and never uses contraceptives (plus, if female, who breast-fed all her children until they weaned themselves) has any platform but hypocrisy from which to oppose future reproductive technology on the grounds of its being unnatural.

Clearly genuflection to nature is no position from which to oppose the future of reproduction.

Gene Pool on the Rocks? Better Shaken than Steered

Suddenly everybody is very worried about the human gene pool. The near-sighted are worried about eugenics. The long-sighted are worried about a reduction in total variation or even the eradication of whole traits. And the concern for all is that if technology has its way, the human gene pool will end up on the rocks, shipwrecked through lack of pilotage. Should we be worried?

In fact, if anything is going to undermine the human gene pool, it will not be cloning or gene therapy. It will be traditional medical science, which for generations has been practising *dysgenics* – its own upside-down form of eugenics. Evidently it is OK to weaken the gene pool but not to enhance it.

Modern medicine uses a variety of treatments that enable many people to survive to old age who might otherwise have died in childhood. Vaccines and antibiotics protect against a broad spectrum of what previously were potentially fatal diseases: polio, scarlet fever, tetanus and whooping cough, for example.

Congenital malformations are surgically repaired. Diabetics and haemophiliacs are injected with essential compounds that they are genetically unable to produce. Childhood cancer and leukaemia are treated with chemicals, irradiation and bone-marrow transplantation.

Counteracting the natural selection effects of disease and genetic deficiencies in these ways creates a gene pool increasingly susceptible to infectious and other diseases in later life. Without the constancy of public and private hygiene and the use of anti-infection agents, epidemics of vast proportions might occur. We are creating a population that is increasingly dependent on medical care. Nobody is likely to complain, of course. We all accept the right of ill people to seek the best help that medical technology can offer. It's not quite so clear who should fund this right, but then we have been here before. It cannot be denied, though – as Marilyn pointed out with regard to impotence in Scene 4 – that medicine will inexorably lead genes that cause their possessors pain and distress to increase in frequency in the gene pool. At least the eugenics of gene therapy would allow medical science to halt and maybe reverse the dysgenic decline it has itself caused.

If people worry about eugenics, and worry about dysgenics, we can only conclude that basically people would like the gene pool to stay much as it is. One thing that would do no harm at all, though, would be to give the gene pool a thorough *shake* to mix it up. Mixing is a *good* thing because it discourages inbreeding, encourages outbreeding and thus breaks up and scatters local clusters of genetic diseases. The ladle that best mixes the gene pool is global migration, which is ironic given that it was the very thing feared most by the racist eugenicists of the past. In the United States, for example, they believed that immigration and interbreeding were increasing genetic defects in the American population.

Wherever we look there has been a recent drop in inbreeding

in human populations. Mating outside the local group is one of the most dramatic and beneficial forces for genetic health in recent evolutionary history. Global migration does little to change the proportion of different genes in the human gene pool. Nor does it eradicate genes. All it does by reducing inbreeding is reduce the chance of hitchhiker genes meeting their match by chance. It thus reduces the number of people who *suffer* from all forms of genetic disease – from mild to extreme – but it does little to change the global gene pool and has no sinister overtones.

In fact, there is only one avenue by which the human gene pool could be threatened by reproduction technology, gene therapy and the Human Genome Project. And that would be if a major body – whether a government, a religion or simply a movement – tried to *steer* the gene pool according to its own prejudices. The feared eradication of homosexuality discussed in Chapter 11 is a case in point. And as we concluded there, even the most exaggerated forms of gene therapy pose little threat *as long as* individuals are granted freedom of choice and are not forced or directed by organizations. We can rely on there being enough variation in human preferences – and enough inaccuracy in genetic knowledge – for there to be no wholesale change in the total gene pool as a result, save the eradication of universally hated genes such as cystic fibrosis.

Paradoxical though it might seem, only pilotage could take the human gene pool on to the rocks.

Human Dignity, the Wills of Gods and Other Irrelevancies

I confess that, as a biologist, I cannot take philosophical and theological arguments about reproductive technology very seriously. Worries about individuals – or even genes – being damaged, victimized or exploited I *can* appreciate. Fears about unleashing disease pandemics or once more snowballing humans towards global overpopulation I can also appreciate. I can even appreciate caution; moratoria on human research while trials are carried out on other animals, even though these often tell us little about the dangers and benefits of techniques for humans. But arguments about human dignity and the wills of Gods leave me cold. It doesn't help that different Gods tend to have different wills, different prophets have different teachings, and different theologians have different interpretations. Nor does it inspire confidence that the ancient prophets whose words are being inter-preted for us had never heard of sperm or genes – or that the interpreters are often celibates.

Anybody who was looking forward at this point to a twenty-page essay on philosophical and theological deliberations on reproductive technology will be disappointed. This section is no longer than, in my view, it merits. In fact, given that dignity defies definition and that I doubt any theologian has a hot line to any God, my own prejudice is that this section is longer than it merits. I intend to give only a flavour of the philosophical and religious rhetoric – esoteric word use that seems little more than an attempt to justify basic gut prejudices.

Much of the highbrow argument surrounding advances in

reproductive technology concerns people's *rights*. And the central question, of course, is whether people have the right to reproduce, no matter what their circumstances. Throughout our evolution, of course, the natural situation meant that people *did not* have a right to reproduce. The situation was simple because there was no alternative. Those that could, did – and those that couldn't, didn't. Rights weren't an issue. Modern technology, however, has created an alternative and society has mooted a right. Now everybody could reproduce if they could only gain access to the technology – or at least, that will be the situation by 2010.

Governments, of course, have the power to deny people such a right. If any given technique were banned it would bar this or that section of the infertile population from reproduction. Ban surrogate mothers, for example, and women who cannot gestate their own child will be doomed never to reproduce. Ban cloning or nucleus transfer and people who cannot manufacture gametes are so denied. But does society itself have the right to deny the pleasures of parenthood to one section of the infertile population but not another? Especially if their rationale for so doing is based on nothing more than tenuous concepts of dignity, divine direction, naturalness – or, on occasion, simply a phobia about unclassifiable individuals with novel pedigrees.

A neat philosophical side-step to the question of rights is the postulation of negative rights. That is, people do have the right to reproduce but society is not obliged to satisfy that right. Such arguments have no immediate resolution and may not seem terribly pertinent to the man or woman who is being denied something that they crave so much – their own, genetic child.

One of the earliest opponents of cloning in the scientific literature implied criticism of what he saw as a fanatical preoccupation with having one's own genetic offspring. People should be satisfied with having a child for the child's sake, it was implied. The child's identity should be shaped by its social and emotional status, not by its genes.

It was a strange and futile claim to make – equivalent to wishing that people had evolved two heads rather than one. Evolutionary theory predicts that all species should care whether the offspring they raise are their genetic products, and empirical data confirm that all species *do* care. Studies of men and women in blended families show graphically and sometimes sinisterly that – consciously or subconsciously – humans also care. *It matters* to them whether a child is 'their' child or somebody else's. The very fact that people demand wherever possible to have their own genetic child shows the depth of feeling involved.

People prefer to raise their own genetic offspring because millennia of natural selection have shaped their psyche to have such a preference. Rarely – as in adoption and fostering – do they fail to show such a preference, and even then it is most often because for medical or pecuniary reasons they have no alternative. Our ancient legacy is to care and to care deeply whether a child is our genetic offspring. It is often inconvenient, but it is the situation. So who has the right to tell anybody that genetic parenthood should not matter?

As an example of the philosophical-cum-ethical logic that reproductive technology can spawn, the following phrases appeared in the pages of *Nature* in the months following the announcement of Dolly, the cloned sheep.

'The moral imperative of preserving human dignity must remain paramount.'

' . . . part of the individuality and dignity of a person probably lies in the uniqueness and unpredictability of his or her development.'

'No man or woman on Earth exists exactly as another has imagined, wished or created. The birth of an infant by cloning would lead to a new category of people whose bodily form and genetic make-up would be exactly as decided by other humans. This would lead to the establishment of an entirely new type of relationship between the "created" and the "creator" which has obvious implications for human dignity.'

And so on . . .

Intellectually challenging though such abstract thoughts about dignity might be to philosophers and theologians, they have little relevance to the childless person desperate to have his or her own genetic child before it is too late. Nor do they have any relevance to the children themselves. If we looked at a class of children, could we pick out those who were conceived by IVF or artificial insemination? Of course not. Unless both were present, could we pick out which child was one half of identical twins? Of course not. Neither could we pick out a clone. All are individuals.

There is a contradiction to the philosophical posturing that has surrounded cloning. On the one hand, groupies to human dignity dismiss genetic determinism, maintaining that the individual is more than the sum of his or her genes. Human dignity transcends genetics, we are told. Twins are never absolutely identical. Yet we are also told that cloning is an affront to human dignity because by controlling the genes the creator is dictating the nature of the created.

The truth, of course, lies somewhere between the two. As we discussed in Chapter 10, a person who cloned him- or herself would be similar to, but not identical to, the clone-child. When the time comes, worried philosophers should comfort themselves by trying to predict a clone's characteristics. Doubtless, if they know the clone-parent, they will get the colour of the eyes and hair, and even the direction of sexual orientation, of the clone right. We can virtually do that for children produced via intercourse already. It will be a surprise, though, if the philosophers can predict with any accuracy the clone's height, weight and attitude – such as how it will feel about dogs (Chapter 6).

Dolly the cloned sheep has become an abstract icon. She is

pitied because of what she is – and even feared and hated because of what she might herald. But she is also an individual, and a character and performer. We know nothing about the six-year-old Finn Dorset sheep that was Dolly's clone-mother. But Dolly herself is a celebrity – and behaves like one. Having spent her early life surrounded by photographers, whenever she sees a camera nowadays she poses as predictably as any Cannes starlet. She walks to the front of her stall, stands on her hind legs and casually drapes her front legs over the top bar of her fence, then keeps her dignified pose until the session is over.

Dolly has now become a mother for the first time. If any philosopher or theologian wants to know what is offensive about their arguments that cloning is an affront to individuality, dignity and the wills of Gods they should watch Dolly for a few minutes. They might decide that of the two it is their argument, not her origin, that is the greater affront to her dignity.

The same will be true for clone-children. Just like little Phoenix in Scene 6, they will be characters with personalities and they will have a place in their parents' and family's affections. The fact that they are clones – or conceived via ICSI, nuclear transfer, frozen embryos, or whatever – will quickly seem irrelevant. They will be as wanted as any other children, filling a place in the parents' lives.

As historical precedence has shown – with IVF, bottle-feeding, contraception and cloning – ethical, philosophical and religious concepts melt away in the face of a real demand by real people. Not necessarily the demand of the majority – at least, not at first – but the demand of a determined minority. And nobody is more determined than a man or a woman who wants to produce their own genetic child but who is prevented from doing so by bad luck, disease, poverty – or red tape.

In the end, reproduction will have no more to do with rights or righteousness than it ever had. People who can, will. Those who can't, won't. The only difference between the future and the past will be that part of being able to reproduce for some people will be gaining access to modern technology. Finances, for example, may become more important than biological fertility. In the Reproduction Restaurant, only conception via intercourse will be on the house. Everything else on the menu will have a price. Some options may be more organic than others – and some will definitely be healthier – but none will be without dignity.

But Where's the Fun?

From the perspective of the twentieth century, a future of contraceptive cafeterias, Reproduction Restaurants, Gamete Marketing Boards and BB schemes probably all seem like bleak prospects. Where are the fun, excitement, passion and emotion that have traditionally gone hand in hand with procreation? By comparison with the past, the technological age that awaits aspiring parents in the future might seem cold and heartless.

But it won't seem like that to its participants. There will be *no* diminution in sexual activity, adventure, emotion or fun. In fact, with the BB scheme in full swing combined with child support legislation and paternity testing, people will be liberated sexually far more than was ever the case in the past. It's just that sex and reproduction will be separated – in reality as well as psychologically.

Future populations will almost certainly be more promiscuous than past ones. Much will depend, though, on how effectively future medicine kills off sexually transmitted diseases. Even if AIDS is defeated, new threats to sexual freedom will inevitably arise and need defeating in their turn. But if medical technology can keep these diseases at bay, reproductive technology will unleash the ancient urges behind human sexuality to a degree that has never before been possible. It will be the 1960s all over again but without the dangers or the attitude. There is no need for the future to be cold and clinical. Our descendants should be able to give their emotions free rein in a way that we never could. The beast within could be released on a longer lead than for centuries.

Despite this, there will doubtless be many people whose ancient urge for space and freedom will drive them to seek environments in which they can act out ancient scenarios far away from GMBs and the like. Doubtless, people will continue to take exotic holidays or, if they can afford it, buy their own Garden of Eden as in the final scene. Doubtless, also, even in the midst of future societies, 1960s-style communes will be set up to denounce all things technological and to try to rediscover ancient roots.

One thing is certain, though. No matter how unnatural it might be and no matter how many esoteric debates are staged in its name, modern technology is here to stay and grow – in reproduction as in all other facets of life.

Even if people do go back to nature from time to time, they will rarely go far from their mobile phone.

Bibliography

Adler, N. A. and J. Schuts, 'Sibling incest offenders', *Child Abuse and Neglect*, 1995, 19: 811–819

Baker, R. R., *Sperm Wars: Infidelity, Sexual Conflict and other Bedroom Battles*, Fourth Estate, 1996

Baker, R. R. and E. R. Oram, *Baby Wars: Parenthood and Family Strife*, Fourth Estate, 1998

Bodmer, W. and R. McKie, *The Book of Man: The Quest to Discover Our Genetic Heritage*, Little, Brown, 1994

Bortolaia Silva, E., Ed., *Good Enough Mothering? Feminist Perspectives on Lone Motherhood*, Routledge, 1996

Buss, D. M., *The Evolution of Desire: Strategies of Human Mating*, Basic Books, 1994

Colborn, T., J. Peterson Myers, et al., *Our Stolen Future*, Dutton, 1996

Daly, M. and M. I. Wilson, 'Child abuse and other risks of not living with both parents', *Ethology and Sociobiology*, 1985, 6: 197–210.

de Waal, F. and F. Lanting, *Bonobo: The Forgotten Ape*, University of California Press, 1997

Ford, C. S. and F. A. Beach, *Patterns of Sexual Behaviour*, Eyre & Spottiswoode, 1952

Gostin, L., Ed., *Surrogate Motherhood: Politics and Privacy*, Indiana University Press, 1990

Hartmann, P. E., S. Rattingan, et al., 'Human lactation: back to nature', *Symposium of the Zoological Society of London*, 1984, 51: 337–368.

Hill, K. and A. Magdalena Hurtado, *Ache Life History: the Ecology and Demography of a Foraging People*, Aldine de Gruyter, 1996

Jones, S., *In the Blood: God, Genes and Destiny*, Flamingo, 1996

Jones, S., R. Martin, et al., Eds., *The Cambridge Encyclopedia of Human Evolution*, Cambridge University Press, 1992

Justice, B. and R. Justice, *The Broken Taboo: Sex in the Family*, Human Sciences Press, 1979

Kevles, D. J., *In the Name of Eugenics*, Harvard University Press, 1995

Kolata, G., *Clone: The Road to Dolly and the Path Ahead*, Allen Lane, The Penguin Press, 1997

Krebs, J. R. and N. B. Davies, *An Introduction to Behavioural Ecology, 3rd Edn*, Blackwell Scientific Publications, 1993

Lyon, J. and P. Gorner, *Altered Fates: Gene Therapy and the Retooling of Human Life*, Norton, 1997

McLanahan, S. and G. Sandefur, *Growing Up with a Single Parent: What Hurts, What Helps*, Harvard University Press, 1994

Pence, G., *Who's Afraid of Human Cloning?*, Rowman & Littlefield, 1997

Riddle, J. M., *Contraception and Abortion from the Ancient World to the Renaissance*, Harvard University Press, 1994

Robertson, J., A. M. Ross, et al., Eds., *DNA in forensic science*, Ellis Horwood, 1990

Rogers, A. R., 'Why menopause?', *Evolutionary Ecology*, 1993, 7: 406–420.

Russell, D. E. H., 'The prevalence and seriousness of incestuous abuse: Stepfathers vs. biological fathers', *Child Abuse and Neglect*, 1984, 8: 15–22.

Silver, L., *Remaking Eden: Cloning and Beyond in a Brave New World*, Avon Books, 1997

Smuts, B. B., D. L. Cheney, et al., *Primate Societies*, University of Chicago Press, 1987

Van Dyck, J., *Manufacturing Babies and Public Consent: Debating the New Reproductive Technologies*, Macmillan, 1994

Wilkie, T., *Perilous Knowledge*, Faber & Faber, 1994

Winston, R., *Getting Pregnant*, Anaya, 1991

Winston, R., *Making Babies*, BBC Books, 1996